HYDROGEN ENERGY: BACKGROUND, SIGNIFICANCE AND FUTURE

HYDROGEN ENERGY: BACKGROUND, SIGNIFICANCE AND FUTURE

ALBERT O. BACKUS

EDITOR

Novinka Books

An imprint of Nova Science Publishers, Inc.

New York

NOTICE TO THE READER

Library of Congress Cataloging-in-Publication Data:
Available Upon Request

ISBN: 1-59454-733-5

Published by Nova Science Publishers, Inc. ✦
New York

CONTENTS

PREFACE

In a world faced with massive and ever-growing energy demands where the environment is an important factor, an array of new technologies and or fuels are becoming a necessity, not a choice. Hydrogen may be one of those alternatives. Hydrogen can be produced from diverse domestic feedstocks using a variety of process technologies. Hydrogen-containing compounds such as fossil fuels, biomass or even water can be a source of hydrogen. Thermochemical processes can be used to produce hydrogen from biomass and from fossil fuels such as coal, natural gas and petroleum. Power generated from sunlight, wind and nuclear sources can be used to produce hydrogen electrolytically. Sunlight alone can also drive photolytic production of hydrogen from water, using advanced photoelectrochemical and photobiological processes.

In: Hydrogen Energy
Editor: A.O. Backus, pp. 1-7

ISBN 1-59454-733-5
© 2006 Nova Science Publishers, Inc.

Chapter 1

HYDROGEN PROGRAM

United States Department of Energy

I. INTRODUCTION

In his 2003 State of the Union Address, President Bush commuicated his vision to the American people that, "the first car driven by a child born today could be powered by hydrogen, and pollution-free." In the 2 years since President Bush launched the Hydrogen Fuel Initiative, the Department of Energy Hydrogen Program staff have worked tirelessly to make this vision a reality and to progress towards a clean and secure energy future.

Congress and Administration Continue to Strongly Support the Hydrogen Economy

Congress has done its part by appropriating 65% more funding in Fiscal Year 2004—$160 million for the first year of the Initiative—than in Fiscal Year 2003 for hydrogen and polymer fuel cell technology. The Senate, under the leadership of Senators Byron Dorgan and Lindsey Graham, have formed a Senate Hydrogen and Fuel Cell Caucus to bring attention to hydrogen's role in long-term national energy and environmental security.

The Administration continues to deliver on President Bush's commitment of $1.2 billion over 5 years by requesting a Fiscal Year 2005 budget of $228 million, a 40% increase over the Fiscal Year 2004 appropriations. The request totals approximately $320 million when hybrid vehicle technologies under FreedomCAR are included. The Administration committed a combined $1.7 billion over 5 years for research under the Hydrogen Fuel Initiative and the FreedomCAR hybrid program.

National Research Council (NRC) Provides Strategic Direction

"A transition to hydrogen as a major fuel in the next 50 years could fundamentally transform the U.S. energy system, creating opportunities to increase energy security through the use of a variety of domestic energy sources for hydrogen production while reducing environmental impacts, including atmospheric CO_2 emissions and criteria pollutants."

-National Research Council Report, "The Hydrogen Economy: Opportunities, Costs, Barriers, and RandD Needs"

At the very beginning of the President's Initiative, the Department asked the NRC to advise us in our planning activities. Since receiving approximately 50 recommendations, the Department has made substantial changes. First, DOE's basic science office has become a formal part of the

Initiative. DOE requested $30 million in Fiscal Year 2005 for basic science to explore the underlying science in hydrogen production, storage and use. In addition, universities have become more dominant players in our new applied research projects in hydrogen storage and renewable production announced by Secretary Abraham.

Second, we have established an independent systems analyses and integration capability at the National Renewable Energy Laboratory (NREL). Richard Truly, NREL Director, has personally overseen establishment of this critical program management function and has created a "firewall" between this function and the research at NREL to avoid conflict of interest. In addition, the Department has hired Mr. Frederick Joseck, as the Hydrogen Program's Technology Analyst, to make decisions regarding the viability of various hydrogen pathways based on energy, environmental and economic factors. The Department has also hired Dr. JoAnn Milliken, as the Hydrogen Program's Chief Engineer, to ensure coherence and consistency among various hydrogen subsystems (production, delivery, storage, etc.) so that the entire system meets customer- driven performance and functional requirements. She is also responsible for establishing the technical baseline configuration and for controlling cost and schedule.

> "The effective management of the Department of Energy hydrogen program will be far more challenging than any activity previously undertaken on the civilian energy side of the DOE."
>
> -National Research Council Report, "The Hydrogen Economy: Opportunities, Costs, Barriers, and RandD Needs"

And finally, the Department has increased its effort on distributed hydrogen production (natural gas reforming and electrolysis) for the transition to a hydrogen economy. While the centralized natural gas option has been eliminated because of long-term supply and import concerns, more fundamental research on carbon-free hydrogen production such as solar-driven biological, photoelectrochemical, and thermochemical production has been enhanced and is a major focus of our new university projects.

As shown in the figure to the right, over 85% of the President's request is for research and devlopment. The Department believes that higher risk research is the predominant role that government should currently play in advancing the hydrogen economy. Projects which lie to the right of the research and development continuum, such as learning demonstrations, are a smaller portion of the overall budget and are much more heavily cost-shared (50%) by the private sector.

The Department is grateful to the NRC's hydrogen committee for its strategic insight and candidness in making recommendations that will improve our research program and make the best use of taxpayer resources.

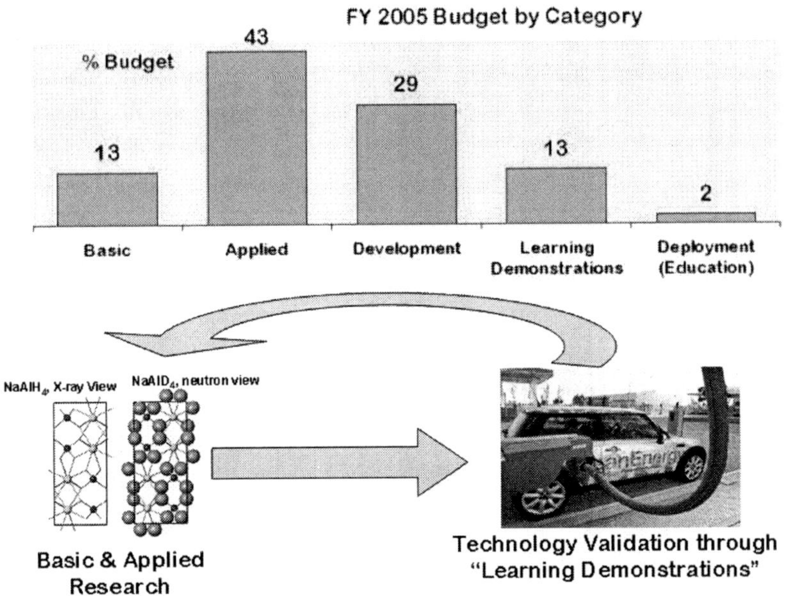

Energy Industry Joins Partnership with Automakers and DOE

In September 2003, DOE expanded its partnership with DaimlerChrysler, Ford and General Motors (through the United States Council for Automotive Research) to include the five major energy companies- BP, ChevronTexaco, ConocoPhillips, ExxonMobil and Shell. The energy companies have become important partners in helping focus DOE-sponsored research to overcome hydrogen infrastructure challenges. Three new technical teams, which consist of scientists and engineers from the 5 energy companies and DOE, were formed in the areas of hydrogen production, delivery and analysis to recommend research priorities and to evaluate technology progress. Hydrogen storage and codes/standards technical teams were expanded and now include members from the energy companies, as well as the automakers and DOE.

Under this partnership, both the energy and automotive industries will provide invaluable assistance to government technology managers by helping establish the technical and economic requirements for satisfying customer needs and for building a business case.

DOE-Sponsored Research Lowers Cost of Hydrogen and Fuel Cells

Air Products and Chemicals, Incorporated, reduced the cost of natural gas-based hydrogen from $5 per gallon gasoline equivalent (gge) in 2003 to $3.60 per gge using innovative reforming and purification technologies. This cost does include co-production of electricity which enables a high capacity factor to be assumed. Further research is needed to achieve the target of less than $2 per gge (untaxed) to be cost-competitive with gasoline.

Las Vegas Power Park Producing Hydrogen for Vehicles and Electricity

National labs and fuel cell developers reduced the high-volume cost of automotive fuel cells from $275/ kW (2002) to $175/kW (2004) using innovative processes for depositing platinum catalyst. Additional Las Vegas

Power Park Producing Hydrogen for Vehicles research is needed for fuel cells to achieve the cost and Electricity target of less than $45/kW to be competitive with gasoline-hybrid engine technologies.

Secretary Abraham Announces Two Rounds of Competitive Grants Worth $425 Million

Using the National Hydrogen Energy Roadmap developed by industry and academia, the Department solicited and selected competitive proposals in the areas of hydrogen production, delivery, storage, fuel cells, education, and for hydrogen fuel cell vehicle/infrastructure "learning demonstrations." With private cost share included, the grants total approximately $675 million. These multi-year projects represent the core research and technology validation to be conducted in the first phase of the President's Hydrogen Fuel Initiative. They involve 60 lead organizations and over 180 partners, including academia, industry and DOE national laboratories. This year's grants are in addition to $75 million of fuel cell grants announced in 2003.

http://www.iphe.net

Secretary Abraham Establishes the International Partnership for the Hydrogen Economy

Secretary Abraham conceived the International Partnership for the Hydrogen Economy (IPHE) and led 15 nations in signing an agreement on November 20, 2003, to formally coordinate research, development and deployment programs that advance the transition to a global hydrogen economy. The IPHE partners represent more than 85 percent of the world's gross domestic product and two-thirds of the world's energy consumption and greenhouse gas emissions. Cooperation is beginning in many areas of technology development and more countries have expressed a strong interest in joining this international partnership.

Department, Government Agencies Better Integrate Hydrogen and Fuel Cell Activities

Together, the DOE renewable, fossil, nuclear and science offices developed an integrated research, development and demonstration plan identifying the key milestones and activities which support an industry commercialization decision in 2015. At a broader level, the White House Office of Science and Technology Policy established a task force among all federal agencies involved in hydrogen and fuel cells. The interagency task force has developed a "taxonomy" of past, present and possible future RandD and a website (www.hydrogen.gov soon to be launched) which will be a repository for all government-sponsored hydrogen and fuel cell activities.

In closing, the past several years have been exciting times as science and technology play a more important role in addressing our Nation's long-term energy and environmental concerns. From hydrogen technology, to FutureGen—a 10-year project to build the world's first coal-fired, emissions-free hydrogen and electricity power plant—to fusion, research in clean energy technology has been a major commitment by the Department of Energy. Since many of these concerns are global, we thank Secretary Abraham for his leadership in establishing the Carbon Sequestration Leadership Forum and the International Partnership for the Hydrogen Economy.

Finally, we are pleased to present the U.S. Department of Energy's Hydrogen Program Annual Progress Report for Fiscal Year 2004. This report documents the progress made on 162 projects currently being conducted. We are also interested in your comments on any projects and in your suggestions on how to better utilize taxpayer resources in advancing the hydrogen economy.

Steve Chalk, Program Manager
DOE Hydrogen Program

In: Hydrogen Energy
Editor: A.O. Backus, pp. 9-73

ISBN 1-59454-733-5
© 2006 Nova Science Publishers, Inc.

Chapter 2

NATIONAL HYDROGEN ENERGY ROADMAP

United States Department of Energy

Based on the results of the National Hydrogen Energy Roadmap
Workshop Washington, DC
April 2-3, 2002

November 2002
. . United States Department of Energy

As we act on President Bush's National Energy Policy, we are focusing on next-generation technologies that expand the diversity of America's supply of energy and "leap frog" the status quo. This requires a revolution in how we find, produce, deliver, store, and use energy.

Hydrogen represents a potential solution to America's needs.

To talk about "the hydrogen economy" is to talk about a world that is fundamentally different from the one we know now.

A hydrogen economy will mean a world where our pollution problems are solved and where our need for abundant and affordable energy is secure...and where concerns about dwindling resources are a thing of the past.

At the Department of Energy, we're not just talking about the hydrogen economy.

We're working to make it a reality.

This Roadmap provides a framework that can make a hydrogen economy a reality.

Spencer Abraham
Secretary of Energy

A PLAN FOR ACTION

Hydrogen holds the potential to provide a clean, reliable, and affordable energy supply that can enhance America's economy, environment, and security. This Roadmap provides a blueprint for the coordinated, long-term, public and private efforts required for hydrogen energy development.

In the coming decades, the United States will need new energy supplies and an upgraded energy infrastructure to meet growing demands for electric power and transportation fuels. Hydrogen provides high efficiency, can be produced from a variety of domestically available resources, and offers near-zero emissions of pollutants and greenhouse gases. Developing hydrogen as a major energy carrier, however, will require solutions to many challenges in the areas of infrastructure, technology, and economics.

The U.S. Department of Energy initiated a National Hydrogen Vision and Roadmap process in response to recommendations in the *National Energy Policy*. The first step in that process resulted in publication of the *National Vision of America's Transition to a Hydrogen Economy* (February 2002).[1] This Roadmap represents the next step in that process.

This Roadmap is neither a government research and development plan nor an industrial commercialization plan. Rather, it explores the wide range of activities required to realize hydrogen's potential in solving U.S. energy security, diversity, and environmental needs. It is intended to inspire the organizations that invest in hydrogen energy systems—public and private, State and Federal, businesses and interest groups—to become involved in a coordinated effort to reduce risk, improve performance, decrease cost, and implement a secure, clean, and reliable energy future.

Meeting the challenges associated with the development of a hydrogen economy will demand considerable time and resources. Activities to progress toward that goal need to begin now.

Mike Davis	Frank Balog
Avista Labs	Ford Motor Company

---

[1] The Vision document can be downloaded from www.eren.doe.gov/hydrogen.

Gene Nemanich
Chevron Texaco Technology
Ventures

Art Katsaros
Air Products and Chemicals,
Inc.

Alan Niedzwiecki
Quantum Technologies

Joan Ogden
Princeton University

Jeff Serfass
National Hydrogen Association

The primary challenge to using more hydrogen in our energy systems is the cost of producing, storing, and transporting it.
— National Energy Policy, 2001

EXECUTIVE SUMMARY

An energy economy based on hydrogen could resolve growing concerns about America's energy supply, security, air pollution, and greenhouse gas emissions. Hydrogen offers the long-term potential for an energy system that produces near-zero emissions and is based on domestically available resources. Before hydrogen can achieve its promise, however, stakeholders must work together to overcome an array of technical, economic, and institutional challenges.

Hydrogen has the potential to play a major role in America's future energy system. This Roadmap outlines key issues and challenges in hydrogen energy development and suggests paths that government and industry can take to expand use of hydrogen-based energy.

Major Findings and Conclusions

Widespread use of hydrogen will affect every aspect of the U.S. energy system, from production through end-use. The individual segments of a hydrogen energy system—production, delivery, storage, conversion, and end-use applications—are closely interrelated and interdependent. Design and implementation of a hydrogen economy must carefully consider each of these segments as well as the "whole system."

Production—Government-industry coordination on hydrogen production systems is required to lower overall costs, improve efficiency, and reduce the cost of carbon sequestration. Better techniques are needed for both central-station and distributed hydrogen production. Efforts should focus on improving existing commercial processes such as steam methane reformation, multifuel gasification, and electrolysis. Development should continue on advanced production techniques such as biological methods and nuclear- or solar-powered thermochemical water-splitting.

Delivery—A greatly expanded distributed infrastructure will be needed to support the expected development of hydrogen production, storage, conversion, and applications. Initial efforts should focus on the development of better components for existing delivery systems, such as hydrogen sensors, pipeline materials, compressors, and high-pressure breakaway hoses. Cost, safety, and reliability issues will influence the planning, design, and development of central versus distributed production and delivery. To address the "chicken and egg" (demand/supply) dilemma, demonstrations should test various hydrogen infrastructure components for both central and distributed systems in concert with end-use applications (e.g., fueling stations and power parks).

Storage—Hydrogen storage is a key enabling technology. None of the current technologies satisfy all of the hydrogen storage attributes sought by manufacturers and end users. Government-industry coordination on research and development is needed to lower costs, improve performance, and develop advanced materials. Efforts should focus on improving existing commercial technologies, including compressed hydrogen gas and liquid hydrogen, and exploring higher-risk storage technologies involving advanced materials (such as lightweight metal hydrides and carbon nanotubes).

> This Roadmap outlines key issues and challenges in hydrogen energy development and suggests paths that government and industry can take to expand use of hydrogen-based energy.
> Ultimately, consumer preferences drive the choices made in energy markets, technology development, and public policy.

Conversion—Conversion of hydrogen into useful forms of electric and thermal energy involves use of fuel cells, reciprocating engines, turbines, and process heaters. Research and development are needed to enhance the manufacturing capabilities and lower the cost of fuel cells as well as to develop higher-efficiency, lower-cost reciprocating engines and turbines. Efforts should focus on developing profitable business models for distributed power systems, optimizing fuel cell designs for mobile and stationary applications, and expanding tests of hydrogen-natural gas blending for combustion. Research is required to expand fundamental understanding of advanced materials, electrochemistry, and fuel cell stack interfaces and to explore the fundamental properties of hydrogen combustion.

Applications—Ultimately, consumers should be able to use hydrogen energy for transportation, electric power generation, and portable electronic devices such as mobile phones and laptop computers. Cost and performance issues associated with hydrogen energy systems will need to be addressed in tandem with customer awareness and acceptance. Key consumer demands include safety, convenience, affordability, and environmental friendliness. Efforts should focus on understanding consumer preferences and building them into hydrogen system designs and operations. Opportunities should be identified to use hydrogen systems in facilities for distributed generation, combined heat and power, and vehicle fleets. Supportive energy and environmental policies should be implemented at the Federal, State, and local levels.

> Ultimately, consumer preferences drive the choices made in energy markets, technology development, and public policy.

All individual segments of the hydrogen industry as well as the overall hydrogen energy system must address several cross-cutting challenges. These challenges include insuring safety, building government/industry partnerships for technology demonstration and commercialization, coordinating activities by diverse stakeholders, maintaining a strong research and development program in both fundamental science and technology development, and implementing effective public policies. Two additional cross-cutting areas could become powerful drivers to assist in addressing these challenges: customer education and the development of codes and standards.

Education and outreach—Hydrogen energy development is a complex topic, and people are uncertain about impacts on the environment, public health, safety, and energy security. Ultimately, consumer preferences drive the choices made in energy markets, technology development, and public policy. Informing the public through educational and training materials, science curricula, and public outreach programs will help garner public acceptance for hydrogen-related products and services.

Codes and standards—Uniform codes and standards for the design, manufacture, and operation of hydrogen energy systems, products, and services can dramatically speed the development process from the laboratory to the marketplace. Government-industry coordination can accelerate codes and standards processes, which must also span national boundaries and be accepted by international bodies to achieve global acceptance.

Development of hydrogen energy technologies represents a potential long-term energy solution for America. A coordinated and focused effort is necessary to bring public and private resources to bear on evaluating the costs and benefits of the transition to a hydrogen economy. Next steps will include the development of detailed research and development plans for each of the technology areas listed above. A significant commitment and coordination of resources will be essential to the success of this effort.

1. INTRODUCTION

Expanded use of hydrogen as an energy carrier for America could help address concerns about energy security, global climate change, and air quality. Hydrogen can be derived from a variety of domestically available primary sources, including fossil fuels, renewables, and nuclear power. Another key benefit is that the by-products of conversion are generally benign for human health and the environment.

Despite these compelling benefits, realization of a hydrogen economy faces multiple challenges. Unlike gasoline and natural gas, hydrogen has no existing, large-scale supporting infrastructure—and building one will require major investment. Although hydrogen production, storage, and delivery technologies are currently in commercial use by the chemical and refining industries, existing hydrogen storage and conversion technologies are still too costly for widespread use in energy applications. Finally, existing energy policies do not promote consideration of the external environmental and security costs of energy that would encourage wider use of hydrogen. Table 1 (below) summarizes the key drivers that support and inhibit the development of hydrogen energy.

Developing hydrogen as a realistic energy option will necessitate an unprecedented level of sustained and coordinated activities by diverse stakeholders. Recognizing the need to develop a coordinated national agenda, the U.S. Department of Energy initiated a National Hydrogen Vision and Roadmap process to incorporate the opinions and viewpoints of a broad cross-section of those stakeholders. The process involved two key meetings: the National Hydrogen Vision Meeting and the National Hydrogen Energy Roadmap Workshop.

Summary of Key Drivers Affecting Hydrogen Energy Development

Support	Inhibit	Both Support and Inhibit
• National security and the need to reduce oil imports • Global climate change and the need to reduce and ultimately stabilize greenhouse gas emissions and pollution • Global population and economic growth • The need for new, clean energy supplies at affordable prices • Air quality and the need to reduce emissions from vehicles and power plants	• The difficulties in building and sustaining national consensus on long term energy policy priorities • Lack of a hydrogen infrastructure and the substantial costs of building one • Lack of commercially available, low-cost hydrogen production[1], storage, and conversion devices • Hydrogen safety issues • The need for additional demonstrations of carbon sequestration and lower-cost sequestration methods	• Rapid pace of technology developments supporting hydrogen and competing energy carriers • The current availability of relatively low-cost fossil fuels exacerbating the inevitable depletion of these resources • Simultaneous consumer preferences for both a clean environment and affordable energy supplies

National Hydrogen Vision Meeting

The National Hydrogen Vision Meeting was held on November 15-16, 2001, in Washington, DC. Participants included more than 50 business executives and public policy leaders from Federal and State agencies, the U.S. Congress, and environmental organizations. The U.S. Department of Energy initiated the meeting in response to recommendations in the National Energy Policy regarding hydrogen technologies. The aims of the meeting were to identify a common vision for the hydrogen economy, the time frame in which such a vision could be expected to occur, and the key milestones for achieving it.[2]

---

[1] While large quantities of hydrogen are currently produced at reasonable cost by steam reformation of methane, this source relies on a limited fossil resource (natural gas) and releases greenhouse gas (CO_2)

[2] Proceedings from the meeting can be downloaded from www.eren.doe.gov/hydrogen.

Vision for the Hydrogen Economy

Hydrogen is America's clean energy choice. Hydrogen is flexible, affordable, safe, domestically produced, used in all sectors of the economy, and in all regions of the country.

Major findings from the vision meeting include the following:

- Hydrogen energy could play an increasingly important role in America's energy future, as it has the potential to help reduce dependence on petroleum imports and lower pollution and greenhouse gas emissions.
- The transition to a hydrogen economy has begun, and could take several decades to achieve.
- The development of hydrogen technologies needs to be accelerated.
- There are "chicken-and-egg" issues regarding market segment development and how supply and demand will push or pull these activities.

Participants at the meeting drew the following conclusions about the vision:

- Federal and State governments will need to implement and sustain consistent energy policies that elevate hydrogen as a priority.
- Strong public-private partnerships will need to focus on finding new ways to collaborate on the development and use of hydrogen energy.
- A logical next step will be the development of a *National Hydrogen Energy Roadmap*, which will need to address research, development, testing, outreach, and codes and standards related to the production, delivery, storage, and use of hydrogen.

The vision document, *"A National Vision of America's Transition to a Hydrogen Economy – To 2030 and Beyond"* outlines the characteristics of a hydrogen economy, explains that the full transition process could take several decades, and describes the potential benefits that a successful transition will produce.[3]

---

[3] The Vision document can be downloaded from www.eren.doe.gov/hydrogen.

National Hydrogen Energy Roadmap Workshop

The National Hydrogen Energy Roadmap Workshop took place on April 2 –3, 2002, in Washington, D.C.[4] Approximately 220 technical experts and industry practitioners from public and private organizations participated in the meeting (a list of participating organizations is located in the Appendix). Seven leaders from industry and academia with expertise in hydrogen systems helped guide the subsequent roadmap development process.

During the workshop, participants divided into breakout groups based on the roadmap segments. They discussed key needs that should be addressed in order to achieve the Vision; appropriate roles for industry, government, universities, and National Laboratories; development of public-private partnerships; and time frames for the activities.

Roadmap Leaders

Leader	Affiliation	Roadmap Segment
Frank Balog	Ford Motor Company	Applications
Mike Davis	Avista Labs	Energy Conversion
Art Katsaros	Air Products and Chemicals, Inc.	Delivery
Gene Nemanich	ChevronTexaco Technology Ventures	Production
Alan Niedzwiecki	Quantum Technologies	Storage
Joan Ogden	Princeton University	Systems Integration
Jeff Serfass	National Hydrogen Association	Public Education and Outreach

Roadmap leaders participated in a panel discussion during the opening plenary session of the Hydrogen Energy Roadmap Workshop on April 2, 2002. Panelists included (from left to right) Mike Davis, Art Katsaros, Frank Balog, Alan Niedzwiecki, Gene Nemanich, Jeff Serfass, and Joan Ogden.

---

[4] A proceedings from the meeting can be downloaded from www.eren.doe.gov/hydrogen.

This Roadmap is a product of the workshop. It is intended to help identify the strategic goals, barriers, and key activities required to evaluate the costs and benefits of a hydrogen economy and to lay a foundation for the public-private partnerships needed to implement the plan. It has been prepared in conjunction with a parallel effort by the U.S.

Department of Energy to report to Congress on "…the technical and economic barriers to the commercial use of fuel cells in transportation, portable power, and stationary and distributed power applications by 2012". The fuel cell report and hydrogen energy roadmap are complementary activities.[5]

The following chapters reflect the ideas and priorities put forth by the workshop participants. Each chapter is focused on an industry segment and provides a description of current status, challenges to achieving the vision, and paths forward.

2. SYSTEMS INTEGRATION

Effective design and implementation of a hydrogen-based energy system requires a "whole system" approach. Complex dependencies among the diverse system components dictate that cross-cutting, system-level issues and concerns receive close attention.

A number of cross-cutting issues will influence hydrogen production, storage, delivery, conversion, applications, education, and outreach:

- Development of national and international codes and standards for hydrogen use (see box on the next page)
- Safety precautions
- Consumer acceptance—providing the expected performance at a reasonable cost
- Collaborative research and development
- Technology validation through demonstrations by government/industry partnerships (The government has a role as an early adopter of integrated hydrogen supply and end-use technologies and as a developer of the hydrogen infrastructure.)

---

[5] The U.S. Department of Energy Fuel Cell Report to Congress—Interim Assessment, April 2002, can be downloaded from www.eren.doe.gov/hydrogen.

- Systems analyses to explore various pathways to widespread hydrogen energy use, including full cost accounting for all competing energy systems
- Ready accessibility to existing information on hydrogen technologies

System integration addresses ways in which different parts of a system work together from technical, economic, and societal standpoints. In many cases, system optimization may require a distinctly different approach from the optimization of a single part. Similarly, a systems focus makes it easier to identify key technical or market barriers in any one part of the system that might impede the development of the whole.

Optimization at the system level will require the following:

- Coordination of technology development between hydrogen producers and end-users (who require hydrogen at a particular purity and pressure)
- A strong, coordinated, and focused research and development program—breakthroughs in hydrogen storage, production, and use could influence how fast and in what way(s) a hydrogen economy develops
- Efficient coordination of supply and demand to solve the perceived "chicken and egg" problem in transportation markets—vehicle manufacturers wish to be assured of fuel supply, while suppliers wish to be assured of a market

Finally, moving to widespread use of hydrogen as an energy carrier involves profound changes in how we view and use energy as individuals and as a society. Actions in the following areas are essential to establishing the underlying set of "system level" preconditions or the context for creating a hydrogen energy system:

- Government leadership to identify and sustain the required long-term activities
- Adoption of policies that consistently incorporate the external costs of energy (such as energy supply security, air quality, and global climate change) and provide a clear signal to industry and consumers

- Development of domestic and international markets for hydrogen energy to harness the projected growth of energy demand in developing nations over the next half century

The ensuing chapters elaborate upon these issues.

Codes and Standards

Impact on Technology Acceptance and Commercialization

Applicable codes and standards are an important enabler for the commercialization of any new technology or product. Although industry uses hydrogen extensively as a chemical, hydrogen use in consumer products will require a completely separate set of codes and standards. The Hydrogen Codes and Standards Coordinating Committee was established to coordinate the diverse activities by the large number of organizations and activities involved in developing and adopting codes for hydrogen technologies. The committee communicates across the hydrogen community and works for the development of consistent codes and standards to accelerate the commercialization of fuel cell and hydrogen technologies.

Codes

The path by which new technologies gain approval elucidates the importance of codes. When codes do not specifically and prescriptively address new technologies (or products, designs, or materials), regulatory authorities rely on information provided by the designer, technology purveyor, or testing entity to validate that the proposed technology/design meets the intent of the code—including all safety requirements. Regulatory authorities unfamiliar with a new product may request substantial documentation, which can drastically delay approvals and could affect competitive positions in the market. To facilitate the introduction of hydrogen as an energy carrier, code officials, industry, and National Laboratories have been working over the past three years to draft new model codes that specifically cover emerging hydrogen technologies for consideration by the various code-enforcing jurisdictions.

The International Code Council is in the process of adopting a new edition of its family of model national uniform building codes. Previous editions of these model codes did not address hydrogen as an energy carrier, nor did they address fuel cells as generating devices or appliances. To remedy these omissions, an Ad Hoc Committee on hydrogen technologies was formed to develop and propose amendments to the codes. Adoption of the proposed amendments by the International Code Council will greatly reduce the time required to include hydrogen technologies in local building codes. Similar efforts are underway with the National Fire Protection Association, which is also in the process of adopting model building codes. Technical experts from industry, universities, and National Laboratories are working with the National Fire Protection Association to ensure that its model codes for hydrogen technologies are consistent with those of the International Code Council.

Standards

The International Standards Organization Technical Committee 197 has also been working to adopt international standards for hydrogen technologies. This international forum has succeeded in getting four standards adopted under the International Standards Organization process, which will be essential in achieving harmonious global standards and regulations for hydrogen applications. Ongoing efforts to establish standards are focusing on establishing safe handling practices, facilitating standard interfaces, eliminating barriers to international trade, and developing quality criteria and testing methods.

3. PRODUCTION

Introduction

Hydrogen can be produced from a variety of sources, including fossil fuels; renewable sources such as wind, solar, or biomass; nuclear or solar heat-powered thermochemical reactions; and solar photolysis or biological methods.

Hydrogen production today: The United States hydrogen industry currently produces nine million tons of hydrogen per year for use in chemicals production, petroleum refining, metals treating, and electrical applications. Hydrogen is primarily used as a feedstock, intermediate chemical, or, on a much smaller scale, a specialty chemical. Only a small portion of the hydrogen produced today is used as an energy carrier, most notably by the National Aeronautics and Space Administration.

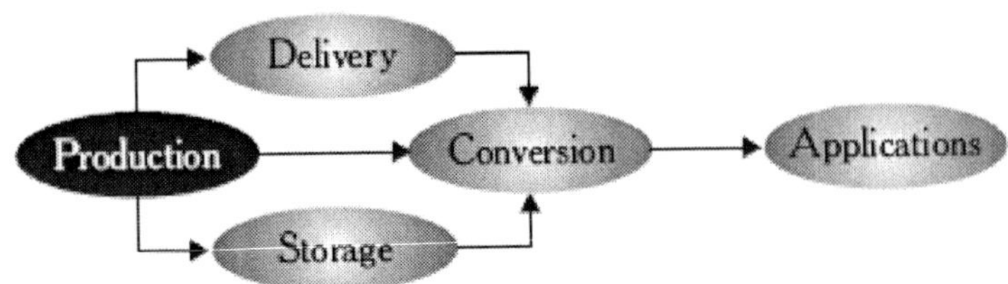

Hydrogen can be produced in centralized facilities or at decentralized locations where it will be used onsite. From centralized facilities, it is distributed to an energy conversion device via pipeline, or stored and shipped via rail or truck. When produced onsite, hydrogen can be stored and/or fed directly into conversion devices for stationary, mobile, and portable applications.

Although hydrogen is the most abundant element in the universe, it does not naturally exist in large quantities or high concentrations on Earth—it must be produced from other compounds such as water, biomass, or fossil fuels. Various methods of production have unique needs in terms of energy sources (e.g., heat, light, electricity) and generate unique by-products or emissions.

Steam methane reforming accounts for 95 percent of the hydrogen produced in the United States. This is a catalytic process that involves reacting natural gas or other light hydrocarbons with steam to produce a mixture of hydrogen and carbon dioxide. The mixture is then separated to produce high-purity hydrogen. This method is the most energy-efficient commercialized technology currently available, and is most cost-effective when applied to large, constant loads.

Partial oxidation of fossil fuels in large gasifiers is another method of thermal hydrogen production. It involves the reaction of a fuel with a limited supply of oxygen to produce a hydrogen mixture, which is then purified. Partial oxidation can be applied to a wide range of hydrocarbon feedstocks, including natural gas, heavy oils, solid biomass, and coal. Its primary by-product is carbon dioxide.

Hydrogen can also be produced by using electricity in electrolyzers to extract hydrogen from water.

Currently this method is not as efficient or cost effective as using fossil fuels in steam methane reforming and partial oxidation, but it would allow for more distributed hydrogen generation and open possibilities for using electricity made from renewable and nuclear resources. The primary by-products are oxygen from the electrolyzer and carbon dioxide from electricity generation.

How much is 9 million tons of hydrogen per year?

Enough to fuel 20 to 30 million hydrogen fueled cars, or enough to power 5 to 8 million homes.

Other methods hold the promise of producing hydrogen without carbon dioxide emissions, but all of these are still in early development phases. They include thermochemical water-splitting using nuclear and solar heat, photolytic (solar) processes using solid state techniques (photoelectrochemical electrolysis), fossil fuel hydrogen production with carbon sequestration, and biological techniques (algae and bacteria).

With most hydrogen now being produced from hydrocarbons, the cost per unit of energy delivered through hydrogen is higher than the cost of the same unit of energy from the hydrocarbon itself.

Vision of hydrogen production: Hydrogen will become a premier energy carrier, reducing U.S. dependence on imported petroleum, diversifying energy sources, and reducing pollution and greenhouse gas emissions. It will be produced in large refineries in industrial areas, power parks and fueling stations in communities, distributed facilities in rural areas, and on-site at customers' premises. Thermal, electric, and photolytic processes will use fossil fuels, biomass, or water as feedstocks and release little or no carbon dioxide into the atmosphere.

A pathway for scaling up hydrogen use would build from the existing hydrogen industry. To foster the initial growth of distributed markets, small reformers and electrolyzers will provide hydrogen for small fleets of fuel cell-powered vehicles and distributed power supply. The next stage of development will include mid-sized community systems and large, centralized hydrogen production facilities with fully developed truck delivery systems for short distances and pipeline delivery for longer distances. As markets grow, costs will drop through economies of scale and technological advances; carbon emissions will decrease with commercialization of carbon capture, sequestration, and advanced direct conversion methods using photolytic, renewable, and nuclear technologies.

Challenges

Multiple challenges must be overcome to achieve the vision of secure, abundant, inexpensive, and clean hydrogen production with low carbon emissions.

Hydrogen Production Costs are High Relative to Conventional Fuels

With most hydrogen currently produced from hydrocarbons, the cost per unit of energy delivered through hydrogen is higher than the cost of the same unit of energy from the hydrocarbon itself. This is especially the case when the comparison is made at the point of sale to the customer, as delivery costs for hydrogen are also higher than for hydrocarbons. The large-scale, well-developed production and delivery infrastructures for natural gas, oil, coal, and electricity keep energy prices low and set a tough price point for hydrogen to meet.

Low Demand Inhibits Development of Production Capacity

Although there is a healthy, growing market for hydrogen in refineries and chemical plants, there is little demand for hydrogen as an energy carrier. Demand growth will depend on the development and implementation of hydrogen storage and conversion devices, and on a demand pull from products such as hydrogen-powered cars and electric generators. Without demand for high-quality hydrogen in the merchant energy carrier market, there is little incentive for industry to completely develop, optimize, and implement existing and new technologies.

Current Technologies Produce Large Quantities of Carbon Dioxide and Are Not Optimized for Making Hydrogen as an Energy Carrier

Existing production technologies can produce vast amounts of hydrogen from hydrocarbons but emit large amounts of carbon dioxide into the atmosphere. Existing commercial production methods (such as steam methane reformation, multi-fuel gasification, and electrolysis) require technical improvements to reduce costs, improve efficiencies, and produce inexpensive, high-purity hydrogen with little or no carbon emissions.

Advanced Hydrogen Production Methods Need Development

While wind, solar, and geothermal resources can produce hydrogen electrolytically, and biomass can produce hydrogen directly, other advanced methods for producing hydrogen from renewable and sustainable energy sources without generating carbon dioxide are still in early research and development phases. Processes such as nuclear thermo-chemical water splitting, photoelectrochemical electrolysis, and biological methods require long-term, focused efforts to move toward commercial readiness. Renewable technologies such as solar, wind, and geothermal need further development for hydrogen production to be more cost-competitive from these sources.

Public-Private Production Demonstrations Are Essential

Stakeholders need a basic understanding of the different sources of hydrogen production before they will be willing to embrace the concepts. Demonstrations are the best way to gain the needed confidence. The large scale of some production processes, however, makes them particularly difficult and expensive to demonstrate.

Paths Forward

The specific needs and actions required to address these barriers differ for each of the hydrogen production technologies. No single technology meets all of the criteria of the Vision; various combinations of the production technologies are likely to be used for different applications.

Enact Policies that Foster Both Technology and Market Development

Government support for research and development should focus on developing advanced renewable and low-carbon-emitting methods plus carbon dioxide capture and sequestration technologies.

Improve Gas Separation and Purification Processes

The oxygen plant is one of the higher cost items in multi-fuel gasifiers; lowering this cost will improve the economics of hydrogen production. Small, low-cost, high-efficiency hydrogen purification methods are needed for distributed reformers that can generate hydrogen at residences or car-refueling stations. Although some purification technologies work well at large commercial sites, these are often difficult to scale down to the size needed for distributed generation.

Develop and Demonstrate Small Reformers

Small reformers that run on natural gas, propane, methanol or diesel can provide hydrogen to some of the first fleets and retail sales points, reducing overall costs. The technology also needs further refinement for improved reliability, longer catalyst life, and integration with storage systems and fuel cells.

Optimize and Reduce Costs of Electrolyzers

Efforts to improve the efficiency and lower the costs of electrolyzers must continue, as this production method is ideal for distributed generation and could offer early market opportunities. Although electrolysis is currently more expensive than thermal production, a better understanding of high-temperature and high-pressure electrolysis could bring costs down. In distributed hydrogen systems, the hydrogen produced on-site often requires compression (to pressures as high as 5,000 psi) for storage; high-pressure electrolysis could remove the need for this additional compression.

A near-term study should be conducted to develop measurable goals for electrolysis in terms of production efficiency, capital cost, and price. Specific goals will help to align and focus development efforts.

Develop Advanced Renewable Energy Methods that Do Not Emit Carbon Dioxide

Photolytic processes use light energy to split water and produce hydrogen, potentially offering lower costs and higher efficiencies for collecting solar energy. Semiconductors that enable photoelectrochemical splitting of water need to become more efficient and less susceptible to corrosion in water. Biological systems may become a "low-tech" way to provide hydrogen, but they are still in the early stages of development.

Develop Advanced Nuclear Energy Methods to Produce Hydrogen

Research is needed to identify and develop methods for economically producing hydrogen with nuclear energy, which would avoid carbon emissions. Thermochemical water splitting using high-temperature heat from advanced nuclear reactors could be included in future nuclear plant designs.

Develop Methods for Large-Scale Carbon Dioxide Capture and Sequestration.

A cost-effective way to capture and sequester carbon dioxide would facilitate the production of vast quantities of hydrogen with low carbon emissions. Capture systems would need to be engineered into plant designs for steam methane reformers and multi-fuel gasifiers to lower the overall systems costs.

Demonstrate Production Technologies in Tandem with Applications

Demonstrations are expensive, especially since there may be little initial demand for the hydrogen produced. Demonstrations that integrate production technology with other elements of the hydrogen infrastructure, including a market use, will be more cost effective. These demonstrations should highlight safety and other benefits to stimulate market interest.

Demonstrations of hydrogen generation, purification, storage, dispensing, and fuel cell electricity generation should be pursued in the short term in major metropolitan areas. For technologies that need larger-scale testing and demonstration, an industrial-scale testing location should be developed to alleviate difficulties in finding acceptable sites.

The ability to capture and sequester carbon dioxide in a costeffective way could open the door to making vast quantities of hydrogen with low carbon emissions.

Conclusion

Research, development, and demonstrations are needed to improve and expand methods of economically producing hydrogen. Production costs need to be lowered, efficiency improved, and carbon sequestration techniques developed. Better techniques are needed for both central-station and distributed hydrogen production. Efforts should focus on existing commercial processes such as steam methane reforming, multi-fuel gasifiers, and electrolyzers, and on the development of advanced techniques such as biomass pyrolysis and nuclear thermochemical water splitting, photoelectrochemical electrolysis, and biological methods.

How Much Hydrogen Do We Need?

Once applications for hydrogen as an energy carrier have become well established, the United States will require much more hydrogen than it now produces. An estimated 40 million tons of hydrogen will be required anually to fuel about 100 million fuel-cell powered cars, or to provide electricity to about 25 million homes.

Each of the following scenarios could produce 40 million tons per year of hydrogen:

Distributed Generation Production Methods

Electrolysis: 1,000,000 small neighborhood based systems could fuel some of the cars and provide some power needs.

Small reformers: 67,000 hydrogen vehicle refueling stations, which is about one third of the current gasoline stations.

Centralized Production Methods

Coal/biomass gasification plants: 140 plants each about like today's large coal fired plants.

Nuclear water splitting: 100 nuclear plants making only hydrogen Oil and natural gas refinery: 20 plants, each the size of a small oil refinery, using oil and natural gas in multi-fuel gasifiers and reformers.

"A Production Mosaic"

Many factors will affect the choice of production methods, how they will be used, and when they might be demonstrated and commercialized. Visualizing a mosaic of future production methods provides a perspective for the Roadmap. The combination of distributed and centralized production, plus advanced methods that are not yet available, could be combined to create a future industry producing 40 million tons of hydrogen per year. Here is one scenario:

100,000 neighborhood electrolyzers	4 million tons
15,000 small reformers in refueling stations	8 million tons
30 coal/biomass gasification plants	8 million tons
10 nuclear water splitting plants	4 million tons
7 large oil and gas SMR/gasification refineries	16 million tons

4. DELIVERY

Introduction

A key element of the overall hydrogen energy infrastructure is the delivery system that moves the hydrogen from its point of production to an end-use device. Delivery system requirements necessarily vary with the production method and end-use application.

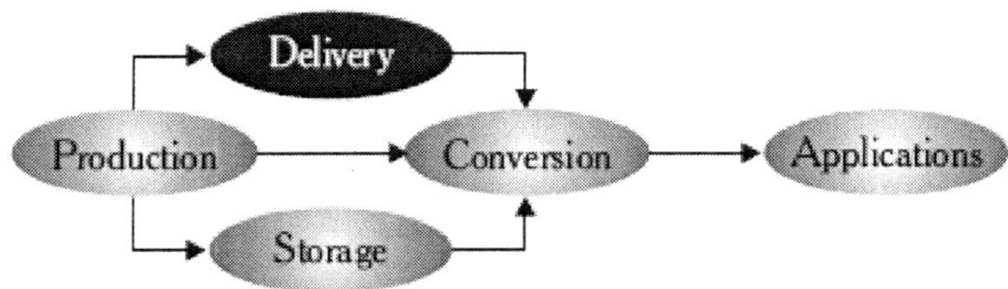

Hydrogen produced in centralized locations is delivered via pipelines, or stored in tubes, tanks, or cylinders that are loaded onto trucks and rail and transported to consumers.

Hydrogen delivery today: At present, hydrogen is produced in a limited number of plants and is used for making chemicals or upgrading fuels. It is currently transported by pipeline or by road via cylinders, tube trailers, and cryogenic tankers, with a small amount shipped by rail car or barge.

As in the case of natural gas distribution, pipelines are employed as an efficient means to supply customer needs. The pipelines are currently limited to a few areas of the United States where large hydrogen refineries and chemical plants are concentrated, such as in Indiana, California, Texas, and Louisiana. The pipelines are owned and operated by merchant hydrogen producers.

Hydrogen distribution via high-pressure cylinders and tube trailers has a range of 100 to 200 miles from the production or distribution facility. For long-distance distribution of up to 1,000 miles, hydrogen is usually transported as a liquid in super-insulated, cryogenic, over-the-road tankers, railcars, and barges and is then vaporized for use at the customer site.

Vision of hydrogen delivery: A national supply network will evolve from the existing fossil fuel-based infrastructure to accommodate both centralized and decentralized production facilities. Pipelines will distribute hydrogen to high-demand areas, and trucks and rail will distribute hydrogen to rural and other lower-demand areas. On-site hydrogen production and distribution facilities will be built where demand is high enough to sustain maintenance of the technologies.

Challenges

A comprehensive delivery infrastructure for hydrogen faces numerous scientific, engineering, environmental, institutional, and market challenges.

An Economic Strategy is Required for the Transition to a Hydrogen Delivery System

Since fueling economics depend on volume, the chicken and egg dilemma (which comes first: fuel or end use applications?) impedes the installation of an effective infrastructure. There is no simple reconciliation between the level of investments required to achieve low costs and the gradual development of the market. Current investments in delivery systems need to be justifiable beyond 2020 to support adequate returns on investment.

Full Life-Cycle Costing Has Not been Applied to Delivery Alternatives

Any strategy to select appropriate delivery systems should involve full life-cycle costing of the options. Life-cycle cost analyses should compare gaseous and liquid hydrogen delivery, and hydrogen carrier media such as metal and chemical hydrides, methanol, and ammonia. Multiple delivery infrastructures may be necessary, which could add to the cost of transitioning to a hydrogen economy.

Hydrogen Delivery Technologies Cost More than Conventional Fuel Delivery

The high cost of hydrogen delivery methods could lead to the use of conventional fuels and associated delivery infrastructure up to the point of use, and small-scale conversion systems to make hydrogen onsite. However, cost effective means do not currently exist to generate hydrogen in small-scale systems.

Current Dispensing Systems are Inconvenient and Expensive

Customers expect the same degree of convenience, cost performance, and safety when dispensing hydrogen fuel as when dispensing conventional fuels. Current hydrogen fueling solutions and designs are not sufficiently mature.

Paths Forward

Current delivery systems will need to expand significantly to deliver hydrogen to all regions of the country in a safe and affordable manner. Distributed hydrogen production is likely to play a significant role, but alternative delivery systems tailored to consumer applications (such as the transport of hydrogen in safe, solid metal alloy hydrides, carbon nanomaterials, and other chemical forms) need to be developed to transport hydrogen to end-use sites on an as-needed basis.

Hydrogen Delivery Challenges

Scientific and Engineering	Environmental and Institutional	Market
Cost-effective means of converting H_2 carriers to H_2 do not exist	Codes and standards do not include H_2 .	Lack of an economic strategy from today's fuels to a hydrogen future
Technical solutions to H_2 dispensing	Lack of harmonization national and international codes	Cost of H_2 technologies higher than current technologies
Proprietary data on materials needed for design are not published	Lack of full social costing of alternatives. Defined value for carbon	Current dispensing system designs do not meet customer expectations for cost, and convenience
Lack of a firm understanding of required H_2 purity for fuel cells	Lack of life-cycle environmental impact to all options	Access to affordable capital
Lack of design criteria for multi-gas pipelines	Liquefaction is energy and greenhouse gases intensive	Current weight and capacity of tube trailers
	Conflicting local vs. state or national interests is a barrier to community acceptance of H_2	Compressed hydrogen has low energy density
	Environmental concerns with fossil carbon-based feedstock	Poor economics for transport of gases over long distance
	Lack of experience and knowledge for operation and maintenance of H_2 technologies	
	Mandates are difficult to establish	

Develop a Demonstration Rollout Plan

A hydrogen delivery infrastructure needs to be started in several regions of the United States. Government-sponsored pilot testing of refueling systems, similar to those for compressed natural gas, would help establish a basis for certifying components of fuel stations. Demonstration programs would stimulate development of delivery and end-use technologies. Regional delivery networks in a number of states would be a good approach to build out the systems.

Develop a Consensus View on Total Costs of Delivery Alternatives

Analyses of the total costs of delivery alternatives need to be conducted. Analyses should weigh options that address all potential fuel delivery points, the cost of maintaining existing fuel infrastructure, and the suitability of the existing infrastructure for future hydrogen use.

Increase Research and Development on Delivery Systems

Improvements are needed in areas such as hydrogen detectors; odorization; materials selection for pipelines, seals, and valves; and transportation containers for hydrogen. Technology validation should address research and development needs for fueling components such as high-pressure, breakaway hoses; hydrogen sensors; compressors; on-site hydrogen generation systems; and robotic fuelers.

Researchers need to test the feasibility of delivery methods from centralized and distributed hydrogen production plants as well as compressors, storage systems, and other components integrated into complete delivery systems.

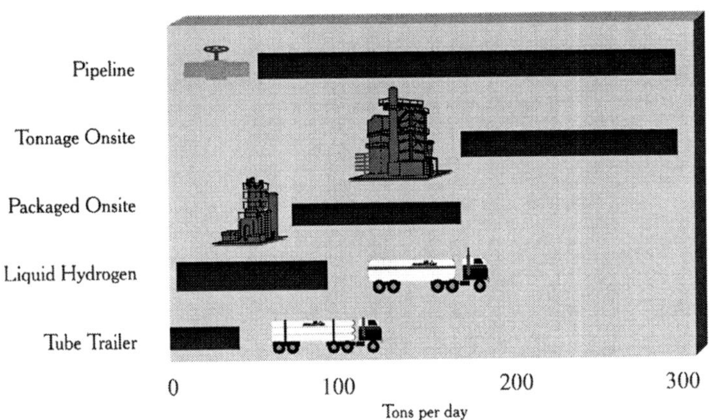

Source: Air Products and Chemicals, Inc.

Hydrogen Delivery Methods

Testing and validation should be ongoing. An organization should be established to perform testing and certification and to identify components that require validation and testing protocols. The organization should include representatives of insurance companies, government agencies, National Laboratories, and industry.

Conclusion

The realization of a hydrogen delivery infrastructure is hampered by the lack of a transition strategy, appropriate codes and standards, the comparatively high cost of hydrogen delivery, and inconvenient dispensing systems.

Installation of hydrogen delivery systems is certainly feasible for industrial markets, which are relatively limited in size and scope. Fuel applications of hydrogen will need to meet much lower economic cost targets than those for current industrial markets. Since scale is critical to cost reductions, care should be taken to build the necessary scale into all government policy and demonstration programs.

Efforts should focus on the development of better components for existing delivery systems, including hydrogen sensors, pipeline materials, compressors, and high-pressure breakaway hoses. To address the "chicken and egg" fuel/use dilemma, demonstration projects should emphasize testing the hydrogen infrastructure components in applications such as fueling stations and power parks.

Hydrogen Delivery - Paths Forward

Needs	Partners
Building codes and equipment standards	Industry with support and funding from government agencies
Multi-state delivery demonstrations and showcases	Industry-led cost-shared partnership with Federal and state governments
Consensus on total costs of fuel alternatives	Government-led and funded with support from National Labs and universities
Improved financial incentives for delivering hydrogen to markets	Federal and state government leadership
Hydrogen economy transition strategy quantified with milestones and targets	Industry- and government-led with support from National Labs and universities

5. STORAGE

Introduction

Storage issues cut across the production, transport, delivery, and end-use applications of hydrogen as an energy carrier. Mobile applications are driving the development of safe, space-efficient, and cost-effective hydrogen storage systems, yet other applications will benefit substantially from all technological advances made for hese onvehicle storage systems.

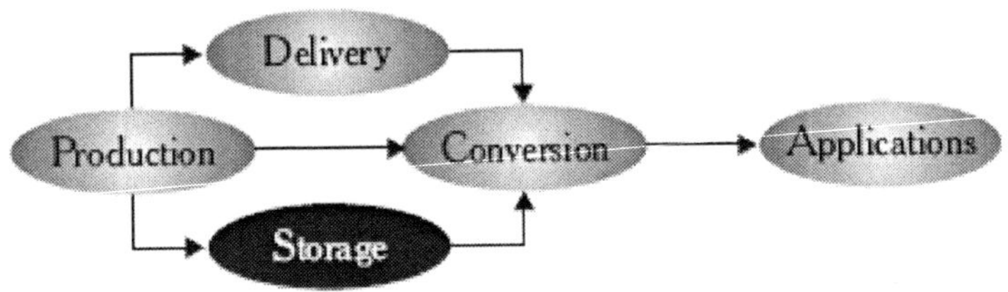

Hydrogen produced in centralized facilities or at decentralized locations may need to be stored before being converted into energy.

Hydrogen storage today: Hydrogen can be stored as a discrete gas or liquid or in a chemical compound. Currently available technologies permit the physical storage, transport, and delivery of gaseous or liquid hydrogen in tanks and pipeline systems. The storage of compressed hydrogen gas in tanks is the most mature technology, though the very low density of hydrogen translates to inefficient use of space aboard a vehicle. This inefficiency can be mitigated with higher compression, such as 5,000 to 10,000 psi. Storage tank designs are advancing with increased strength-to-weight ratio materials and optimized structures that provide better containment, reduced weight and volume, improved impact resistance, and improved safety.

Liquid hydrogen takes up less storage volume than gas but requires cryogenic containers. Furthermore, the liquefaction of hydrogen is an energy-intensive process and results in large evaporative losses—about one-third of the energy content of the hydrogen is lost in the process.

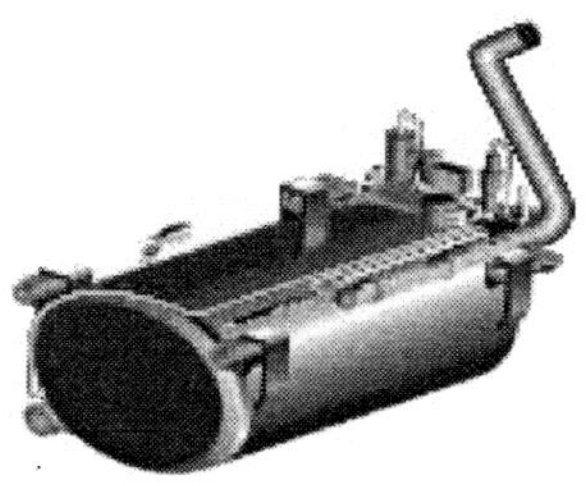

Hydrogen can be stored at high densities as reversible metal hydrides or adsorbed on carbon structures. When the hydrogen is needed, it can be released from these materials under certain temperature and pressure conditions. Complex-based reversible hydrides such as alanates have recently demonstrated improved weight performance over metal hydrides along with modest temperatures for hydrogen recovery. The most promising carbon materials for hydrogen storage at this time appear to be carbon nanotubes.

Chemical hydrides are emerging as another alternative to direct hydrogen storage. The chemical hydrides considered for storage applications are a class of compounds that can be stored in solution as an alkaline liquid. Since the hydrogen is chemically bound in the compound and released by a catalyzed process, chemical hydrides present an inherently safer option than the storage of volatile and flammable fuel, be it hydrogen, gasoline, methanol, etc. The challenges associated with chemical hydrides include lowering the cost of the "round-trip" chemical hydride process (which requires recycling of spent "fuel"), increasing overall "well to wheels" energy efficiency, and development of infrastructure to support the production, delivery, and recycling of the chemical hydrides for transportation and other uses.

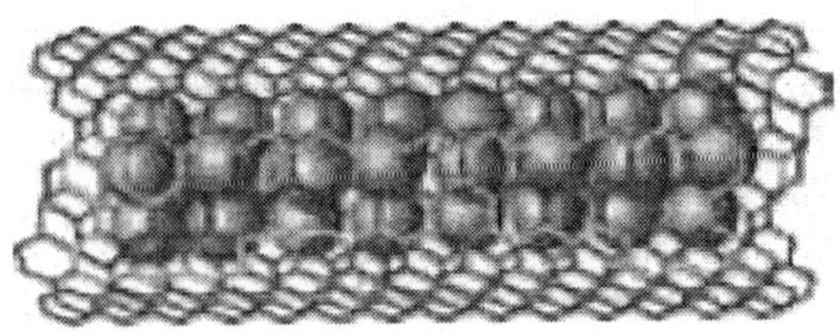

No current technology appears to satisfy all of the desired storage criteria sought by manufacturers and end users. Compressed hydrogen storage is a mature technology, though improvements in cost, weight, and volume storage efficiency must continue to be made. Several automotive manufacturers are considering liquid hydrogen storage because of its good volumetric storage efficiency; however the special handling requirements, long-term storage losses, and cryogenic liquefaction energy demands currently detract from its commercial viability. Metal hydrides offer the advantages of lower pressure storage, conformable shapes, and reasonable volumetric storage efficiency, but have weight penalties and thermal management issues. Although chemical hydrides present a potentially safer and more volumetrically efficient option, there are a number of challenges that must be addressed, including cost, recycling, overall energy efficiency, and infrastructure. Adsorbing materials with high surface areas are emerging, but the design of practical systems awaits a better understanding of the fundamental adsorption/ desorption processes and development of high-volume manufacturing processes for the materials.

Vision of hydrogen storage: A selection of relatively lightweight, low-cost, and low-volume hydrogen storage devices will be available to meet a variety of energy needs. Pocket-sized containers will provide hydrogen for portable telecommunications and computer equipment, small and medium hydrogen containers will be available for vehicles and on-site power systems, and industrial-sized storage devices will be available for power parks and utility-scale systems. Solid-state storage media that use metal hydrides will be in mass production as a mature technology. Storage devices based on carbon structures will be developed.

Compressed Fuel Storage	■ Cylindrical tanks ■ Quasi-conformable tanks
Liquid Hydrogen Storage	■ Cylindrical tanks ■ Elliptical tanks ■ Cryotanks ■ HP liquid tanks
Solid State Conformable Storage	■ Hydride materials ■ Carbon adsorption
Chemical Hydrides	■ Off-board recycling

Hydrogen Storage Alternatives

Challenges

Hydrogen storage must meet a number of challenges before hydrogen can become an acceptable energy option for the consumer. The technology must be made transparent to the end user—similar to today's experience with internal combustion gasoline-powered vehicles. Specific challenges include the following:

Current Research and Development Efforts are Insufficient

Hydrogen storage is a critical enabling element in the hydrogen cycle, from production and delivery to energy conversion and applications. Improved storage technologies are needed to satisfy end-user expectations and foster consumer confidence in hydrogen-powered alternatives. A substantial research and development With transparency an acknowledged target, it investment in hydrogen storage technologies will be required is important to understand consumers' to achieve the performance and cost targets for an acceptable expectations for fuel storage on a vehicle. storage solution.

What Consumers Want
With transparency an acknowledged target, it is important to understand consumers' expectations for fuel storage on a vehicle. Simply put, consumers do not think about fuel storage. They do not see the fuel tank. They expect maximum passenger and trunk space.
They expect 300 to 400 miles range on "a tank of gas" before having to fill-up. They expect to fill up their "tank" in less than 3-5 minutes.

They are used to self-serve "gas stations" that are virtually foolproof, with a simple trigger-type nozzle that starts with a push of a button or flick of a lever. They expect to be able to refuel at the corner gas station, although refueling at home would be a nicety. Probably the only time most consumers in the United States think about fuel, let alone fuel storage, is when fuel prices rise to $2.00 per gallon of gasoline.

New media development is needed to provide reversible, low-temperature, high-density storage of hydrogen. These storage characteristics generally describe the technical goals for some of the solid-state materials, including hydrides and carbon adsorption materials. The ultimate hydrogen storage system for meeting manufacturer, consumer, and end-user expectations would be low in cost and energy efficient, provide fast-fill capability, and offer inherent safety.

Energy storage densities are insufficient to gain market acceptance. This barrier directly relates to making hydrogen storage transparent to the consumer and end user.

Specifically, transparency would mean a hydrogen storage system that enables a vehicle to travel 300 to 400 miles and fits in an envelope that does not compromise either passenger or storage space. Fundamental limitations on hydrogen density will ultimately limit storage performance. The performance of vehicles, therefore, depends on the overall system performance—the combined vehicle efficiency, energy conversion efficiency, and storage efficiency.

Low Demand Means High Costs

As there are few hydrogen-fueled vehicles on the road today, the more mature compressed and liquid hydrogen storage technologies are quite expensive. High-pressure cylinders will be amenable to high-volume production, once demand warrants it. Raw material costs could also be reduced substantially if there were sufficient demand. For emerging technologies, manufacturing feasibility and cost reduction measures will play integral roles in the technology development process. The initially low rates at which automakers expect to introduce fuel cell vehicles will present a challenge to the commercialization and cost reduction of hydrogen storage technologies.

Paths Forward

Achieving the Vision for hydrogen storage will require coordinated activities that address the challenges.

Develop a Coordinated National Program to Advance Hydrogen Storage Materials

A fully funded national program is needed to improve the performance and reduce the cost of hydrogen storage. Emerging hydrogen applications require a range of near-term advances in high-pressure storage technology: improved weight efficiency, service pressure capability, conformable or quasi-conformable shapes, system integration and packaging, "smart" tanks with integrated or embedded sensors, and system costs. Advanced storage materials that show promise for hydrogen storage include alanates, carbon structures, chemical hydrides, and metal hydrides. While promising, these storage technologies are still in the developmental stage. Further research and development efforts are required to understand how to produce and contain the materials, fill and discharge hydrogen from them, manage the pressure and thermal properties, integrate the materials into a practical system, and meet infrastructure requirements.

> - Type 1: all metal cylinder
> - Type 2: load-bearing metal liner hoop wrapped with resin-impregnated continuous filament
> - Type 3: non-load-bearing metal liner axial and hoop wrapped with resin-impregnated continuous filament
> - Type 4: non-load-bearing non-metal liner axial and hoop wrapped with resin-impregnated continuous filament
> - Type 5 (Other): type of construction not covered by Types 1 to 4 above.

Source: European Integrated Hydrogen Program

Hydrogen Storage Tank Classifications

Elevate Research and Development in Hydrogen Storage to a Level Commensurate with its Importance

Storage technologies are integral to the production, transport, delivery, and application of hydrogen as an energy carrier. Since mobile applications impose severe size and weight constraints, they are a major driver behind efforts to develop safe, space-efficient, and cost-effective hydrogen storage systems—but other applications will also benefit from any technological advances made for on-board systems.

Current research and development specifically allocated to advanced hydrogen storage technologies is inadequate to fully investigate, develop, and demonstrate all materials. Research sponsored by the U.S. Department of Energy, the U.S. Department of Transportation, and the U.S. Department of Defense should be coordinated, with similar activities sponsored by industry.

Initiate a Program to Support Development of High-Risk Technologies

None of the currently known technologies satisfy all the desired hydrogen storage attributes sought by manufacturers and end users. Each has its advantages and disadvantages. While improvements in currently known storage technologies are likely, a technology breakthrough may be necessary to achieve an "ultimate" storage device. This is an excellent example of an area of research where there needs to be freedom, and associated funding, to pursue non-obvious technology solutions not currently known and independent of traditional performance metrics. This approach is high-risk, but carries the potential for high payback.

Develop a Mass Production Process for Hydrogen Storage Media

Currently, no market force is driving efforts to reduce raw material costs and develop efficient mass production processes. Even the more mature compressed and liquid hydrogen storage technologies are expensive due to an absence of highvolume demand. Emerging technologies still in the laboratory, including hydrides, alanates, and carbon adsorption materials, have further to go along the path to commercialization and mass production. Fundamental improvements in hydrogen storage processes remain to be fully understood and optimized. Once the materials have been optimized in the laboratory, practical integrated storage systems must be developed and demonstrated. At that point, design and scale-up for production and cost must be addressed.

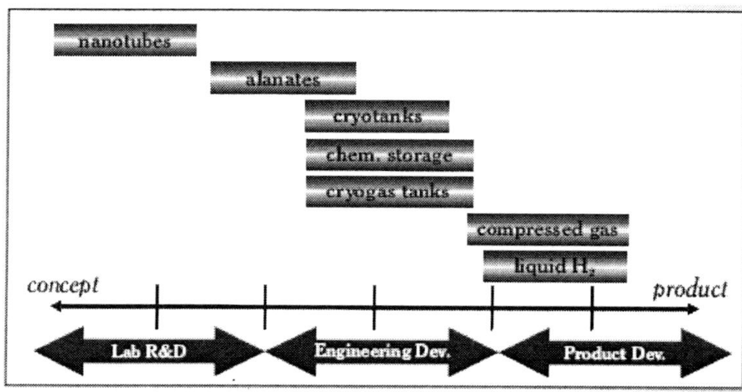

Source: George Thomas, Ph.D.

Storage Technology Status ·

Conclusion

The lack of low-cost and lightweight storage devices as well as commercially available and cost-competitive fuel cells interferes with the implementation of hydrogen as an energy carrier. For a "hydrogen economy" to evolve, consumers will need to have convenient access to hydrogen, and storage devices will be one of the keys. Better hydrogen storage systems will offer easy access to hydrogen for vehicles, distributed energy facilities, or central station power plants.

The most challenging application is the light-duty vehicle or, more specifically, the automobile. Automobiles impose the greatest constraints with respect to available space on-board the vehicle and the greatest consumer expectations for energy density (vehicle range). In the near-term, fuel cell vehicles are likely to be introduced first in fleet applications. Since fleet applications are apt to have centralized refueling facilities, a vehicle range of 100 to 150 miles (160 to 241 kilometers) would be acceptable. In terms of mass of hydrogen, this range could be achieved with about 3 kilograms of hydrogen supplying a fuel cell vehicle. Mature compressed and liquid hydrogen storage technologies of reasonable size and weight could achieve this short-term goal.

In the longer term, average consumers will expect fuel cell vehicles to provide the same cost, convenience, and operational characteristics as gasoline-powered vehicles. In fact, it is likely that fuel cell vehicles will have to offer a significant value proposition to encourage consumers to adopt a new technology rather than continue with something that is tried and true. Vehicle range will be an important factor to consumers, especially as a hydrogen refueling infrastructure begins to develop. Fuel cell vehicle ranges of 300 to 400 miles (480 to 644 kilometers) will be needed, requiring roughly 5 kilograms of hydrogen to be stored on-board. Advanced storage methods, including advancements in compressed storage, alanate hydrides, cryogas tanks, and carbon nanostructures, will have to emerge from the laboratory to reduce hydrogen storage system size, weight, and cost without sacrificing safety or consumer convenience.

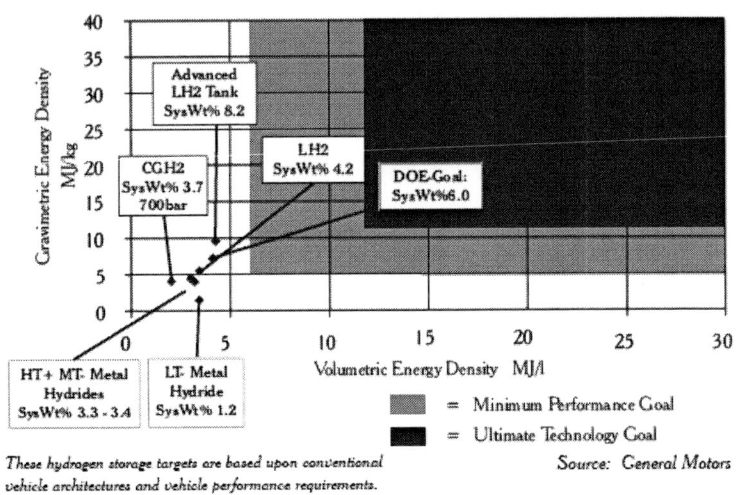

Gravimetric Energy Density vs. Volumetric Energy Density of Fuel Cell Hydrogen Storage Systems

6. CONVERSION

Introduction

Hydrogen can be used both in engines and in fuel cells. Engines can combust hydrogen in the same manner as gasoline or natural gas, while fuel cells use the chemical energy of hydrogen to produce electricity and thermal energy. Since electrochemical reactions are more efficient than combustion at generating energy, fuel cells are more efficient than internal combustion engines.

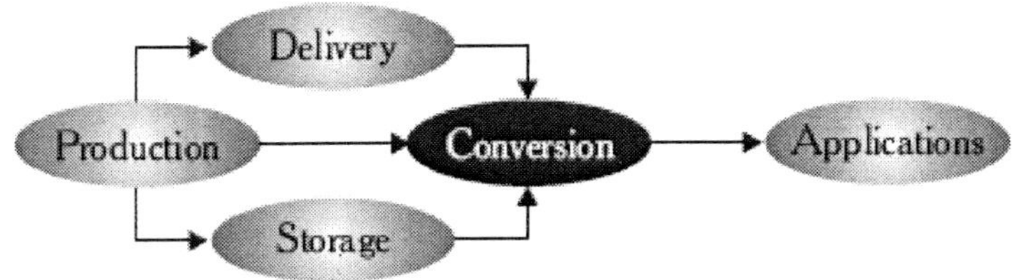

Once hydrogen is produced and delivered to a conversion device, it is used to fuel combustion engines and fuel cells that generate electrical, mechanical, and thermal energy.

Hydrogen conversion today: The use of hydrogen in engines is a fairly well developed technology—the National Aeronautics and Space Administration and the Department of Defense use it in the space shuttle's main engines and unmanned rocket engines. Other combustion applications are under development, including new combustion equipment designed specifically for hydrogen in turbines and engines. Vehicles with hydrogen internal combustion engines are now in the demonstration phase, and the combustion of hydrogen blends is being tested.

Fuel cells are in various stages of development. Current fuel cell efficiencies range from 40 to 50 percent at full power and 60 percent at quarter-power, with up to 80-percent efficiency reported for combined heat and power applications.

Phosphoric-acid fuel cells are the most developed fuel cells for commercial use. Many stationary units have been installed to provide grid support and reliable back-up power, and mobile units are powering buses and other large vehicles.

Polymer-electrolyte membrane fuel cells are being developed and tested for use in transportation, stationary, and portable applications. Interest in polymer-electrolyte membrane fuel cells has experienced a tremendous upsurge over the past few years, and most major automotive manufacturers are developing fuel cell concept cars. Alkaline fuel cells have been used in military applications and space missions (to provide electricity and drinking water for astronauts). Currently they are being tested for transportation applications.

Solid-oxide and molten-carbonate fuel cells are best for use in generating electricity in stationary, combined-cycle applications and cogeneration applications in which waste heat is used for cogeneration. They are also a good fit for portable power and transportation applications, especially large trucks.

Fuel cells can and currently do run on fossil fuel reformate.

Hydrogen Conversion Technologies and Applications

Technology	Application
Combustion	
Gas Turbines	• Distributed power • Combined heat and power • Central station power
Reciprocating Engines	• Vehicles • Distributed power • Combined heat and power
Fuel Cells	
Polymer Electrolyte Membrane (PEM)	• Vehicles • Distributed power • Combined heat and power • Portable power
Alkaline (AFC)	• Vehicles • Distributed power
Phosphoric Acid (PAFC)	• Distributed power • Combined heat and power
Molten Carbonate (MCFC)	• Distributed power • Combined heat and power
Solid Oxide (SOFC)	• Truck APVs • Distributed power • Combined heat and power

Vision of hydrogen energy conversion: Fuel cells will be a mature, cost-competitive technology in mass production. Advanced, hydrogen-powered energy generation devices such as combustion turbines and reciprocating engines will enjoy widespread commercial use.

The commercial production, delivery, and storage of hydrogen will go hand in hand with the commercial conversion of hydrogen into valuable energy services and products, such as electricity and thermal or mechanical energy. As shown in the table above, the technologies appropriate for commercial conversion include established technologies, such as combustion turbines and reciprocating engines, as well as less developed technologies with great potential, such as fuel cells. Current products embodying these technologies have the potential to provide safe, clean, and affordable energy services in all sectors of our global economy.

Challenges

All of today's conversion products, demonstration models, and prototypes possess some deficiencies; they cannot yet provide, at an affordable cost, the level and quality of energy services demanded by a broad base of consumers. While fuel cell technologies have generated much excitement, they are still in various stages of maturity. Most have not been manufactured in large quantities, and numerous performance issues—including durability, reliability, and cost—remain to be resolved. Combustion turbines and engines that use hydrogen or hydrogen/natural gas blends, already in use in both mobile and stationary applications, are much closer to satisfying these criteria than are fuel cells.

Top Priority Hydrogen Conversion Challenges

Scientific	Engineering	Market	Institutional
Knowledge gaps in materials science/technology and electrochemistry	Questions on fuel cell performance and durability	No "value proposition" for using hydrogen rather than fossil fuels (externality costs for fossil fuels not in fuel prices)	Lack of product safety codes for hydrogen using devices
	High fuel cell manufacturing costs		Lack of safety standards for installation and operation of hydrogen conversion systems in vehicle and buildings
Knowledge gaps in hydrogen combustion properties	Unproven hydrogen-burning engine and turbine performance	Lack of profitable business models for widespread installation of distributed energy systems and CHP	
	Questions about flame management and impacts on engine and turbine designs	Large financial risk of being overtaken by competing technologies	Lack of national and state public policies for expanding the use of hydrogen conversion devices

No Single Fuel Cell Technology Has Met All the Basic Criteria for Performance, Durability, and Cost

Basic and applied research in materials science and electrochemistry is required to improve the design and operation of all fuel cell technologies and provide an ongoing basis for substantial cost reduction and performance improvements. In polymer-electrolyte membrane fuel cells, for example, researchers still need a better understanding of the purity levels of graphite; durability of membranes; gas diffusion layers; bipolar plates; and long-term contamination issues. In solid-oxide fuel cells, researchers are working to expand their understanding of interconnected materials and the processes that underlie sealing and joining.

Fuel Cells Require Enhanced Materials, Membranes, and Catalysts to Meet Both Engineering and Cost Criteria

For all types of fuel cells except phosphoric acid, reliability of performance and durability over extended hours of operation remain to be proven. Phosphoric-acid fuel cells are the only type of fuel cells with substantial commercial experience, but efforts to bring down their manufacturing cost have not yet paid off. Questions also remain about the performance of all types of fuel cells under diverse climatic conditions and geographic locations. Manufacturing scale-up issues and the associated need to establish high-volume demand are major barriers in achieving cost reductions.

Research is Needed to Fill in Critical Knowledge Gaps

Researchers require better information about the flame characteristics of hydrogen combustion and the impacts of conversion technologies on reciprocating engine and turbine designs. Better knowledge is also needed to guide the use of advanced materials in hydrogen combustion systems. Existing databases need to be populated with more performance data for hydrogen-burning engines and turbines operating over extended periods; performance data needs include efficiency, emissions, and safety, for both mobile and stationary applications.

Market and Institutional Barriers Hinder Development of Cost-Competitive Hydrogen Conversion Devices

Customers do not see a robust value proposition that convinces them to choose hydrogen conversion products. Substantial cost reduction will be essential—particularly without a bridging incentive or government mandate fostering use of hydrogen conversion products rather than lower-cost conventional fuels and products. In the absence of such policies, conventional fuels and conversion devices will continue to be the only practical option for consumers. Fuel cell manufacturers also face problems in developing innovative safety technologies and achieving profitable operations prior to the development of large-scale markets.

Paths Forward

Over the last decade, most of the enhancements to fuel cell performance have been achieved through incremental improvements to known materials and processes. Investing in efforts to increase fundamental understanding of current materials, interfaces, and processes will support important advances, such as the following:

- Better electrocatalysts to reduce the cathode over-potential and/or tolerate carbon monoxide at the anode
- Non-precious metal catalysts to dramatically reduce cost
- Higher-temperature proton-exchange membrane fuel cells to facilitate thermal management and improve combined heat and power potential
- Lower-temperature oxide-ion conductors to enable moderate-temperature solid-oxide fuel cells and broaden materials choices, thus lowering system costs

Continue Research and Development on Fuel Cells and Combustion Engines

Fundamental research is needed to advance scientific understanding of the materials used in fuel cells, particularly their chemical and physical properties and their interactions. This research will require a three-pronged effort in materials science, interfaces, and electrochemistry. Advancements in these areas could lead to new designs and open the possibility of using lower-cost and easier-to-manufacture materials. Significant advances are needed for stack materials, oxygen cathodes, and membranes.

Researchers need better methods for characterizing materials as well as better understanding of advanced ceramic materials and membrane degradation mechanisms. Improvements in these areas could lead to the development of polymer-electrolyte membrane fuel cells that operate at higher temperatures, and solid-oxide fuel cells that operate at lower temperatures.

Other key research needs include engine and turbine materials that resist corrosion and operate efficiently at higher temperatures, more durable and lower-cost sensors and instrumentation, and better-performing hydrogen-natural gas fuel blends.

Continuing investment in technology development and manufacturing methodologies, starting with components and stacks and continuing through system integration, will hasten commercialization. Industry-driven, cost-shared partnerships with government, supported by universities and laboratories, could lead to catalysts with better performance and lower costs.

Enhance Manufacturing Capabilities for Fuel Cells

Techniques are needed for handling high fuel cell production volumes and achieving better consistency and quality control. Advancements in this area are one of the surest means to achieving the large cost reductions needed to move fuel cells from niche to mass markets. Improvements are also needed in the cost and integration of balance-of-plant components, such as power conditioning, thermal storage and management, water management, and fuel processing equipment.

Top Priority Hydrogen Conversion Needs

Fuel Cells	Combustion	Demonstrations	Codes and Standards	Analysis
Expanded fundamental research program in advanced materials, interfaces and electrochemistry Lower cost designs Enhanced manufacturing capabilities Lower cost balance-of-plant components	Higher efficiency and lower cost engine and turbine designs Instrumentation and controls optimized for hydrogen combustion parameters Analysis of hydrogen-natural gas blending for lower emissions	Expand number of sites to include wider range of technologies, applications, and environmental conditions Expand information dissemination Expand validation of hydrogen combustion	Product safety standards Building codes (fire, safety, plumbing) Vehicle standards Utility interconnection standards	More credible market analysis Catalog existing research results and disseminate widely Software tools to simulate collisions to enhance fuel cell engine designs

Develop New Engine and Turbine Designs to Lower Nitrogen Oxide Emissions when Combusting Hydrogen-Natural Gas Blends

A full range of possibilities for blending hydrogen with natural gas merit further exploration as a low-emissions strategy. New engine and turbine controls should be developed to optimize performance when used in conjunction with hydrogen storage devices. Use of catalytic combustion techniques is another promising area in need of further exploration.

Collect More and Better Information on Operating Performance at Existing Demonstration Sites

Improved instrumentation and expanded data collection efforts are required to facilitate analysis of the full range of cost, efficiency, and emissions parameters for all mobile and stationary applications under a wider range of environmental conditions. More extensive tests of the reliability and durability of advanced materials are also needed, particularly for polymer-electrolyte membrane and solid-oxide fuel cells. At the same time, better market analysis is needed to provide the financial community with an improved understanding of the potential for fuel cells and hydrogen-using engines and gas turbines.

Conclusion

Engines, combustion turbines, and fuel cells can convert hydrogen into useful forms of energy. Research and development are needed to lower costs and enhance manufacturing capabilities for fuel cells and to develop higher-efficiency and lower-cost designs for engines and turbines. Industry should focus its efforts on developing profitable business models for distributed power systems, optimizing fuel cell designs for mobile and stationary applications, and expanding tests of hydrogen-natural gas blends for combustion. Government should assist in developing better information on the fundamental properties of hydrogen combustion and improving fundamental understanding of advanced materials, electrochemistry, and interfaces for fuel cells.

7. APPLICATIONS

Introduction

Hydrogen can be used in conventional power generation technologies, such as automobile engines and power plant turbines, or in fuel cells, which are relatively cleaner and more efficient than conventional technologies. Fuel cells have broad application potential in both transportation and electrical power generation, including on-site generation for individual homes and office buildings.

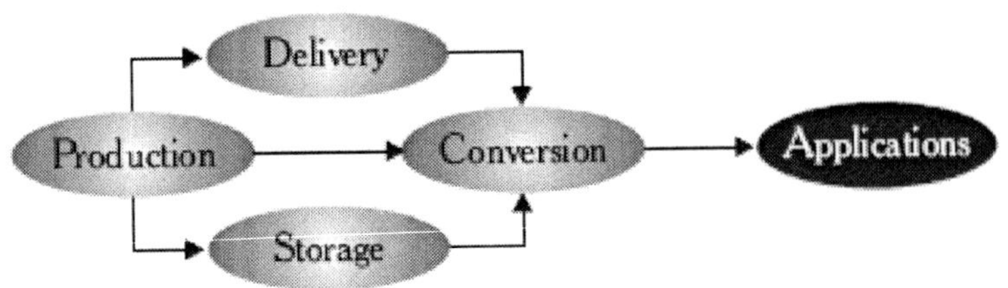

Conversion devices are installed into end-use applications and used to generate electricity for buildings and communities, and to power vehicles and portable devices.

Transportation applications today: Transportation applications for hydrogen include buses, trucks, passenger vehicles, and trains. Technologies are being developed to use hydrogen in both fuel cells and internal combustion engines, including methanol systems.

Nearly every major automaker has a hydrogen-fueled vehicle program, with various targets for demonstration between 2003 and 2006. The early fuel cell demonstration programs will consist of pilot-plant "batch-builds" of approximately 10 to 150 vehicles. These early vehicles are most likely to be deployed in fleets with a centralized or shared re-fueling infrastructure to limit capital investment. Information obtained from these vehicle demonstrations will then be used to help determine how and when to advance to the next level of production.

Hydrogen-fueled internal-combustion engine vehicles are viewed by some as a near-term, lower-cost option that could assist in the development of hydrogen infrastructure and hydrogen storage technology. A key advantage of this option is that hydrogen-fueled internal-combustion engines vehicles can be made in larger numbers when demand warrants.

Stationary power generation today: Stationary power applications include backup power units, grid management, power for remote locations, stand-alone power plants for towns and cities, distributed generation for buildings, and cogeneration (in which excess thermal energy from electricity generation is used for heat). Although commercial fuel cells are on the market, the industry is still in its infancy. Most existing fuel cell systems are being used in commercial settings and operate on reformate from natural gas. Widespread availability of hydrogen would allow the introduction of direct hydrogen units—simpler systems with lower cost and increased reliability.

In general, combustion-based processes, such as gas turbines and reciprocating engines, can be designed to use hydrogen either alone or mixed with natural gas. These technologies tend to have applications in the higher power ranges of stationary generation.

Portable power generation today: Portable applications for fuel cells include consumer electronics, business machinery, and recreational devices. Many participants in the fuel cell industry are developing small-capacity units for a variety of portable and premium power applications ranging from 25-watt systems for portable electronics to 10-kilowatt systems for critical commercial and medical functions. Most of these portable applications will use methanol or hydrogen as the fuel. In addition to consumer applications, portable fuel cells may be well suited for use as auxiliary power units in military applications.

Characteristic Requirements of Transportation vs. Stationary Applications

Characteristic	Transportation	Stationary
Max Power	50 - 100 kW	2 - 200 kW (Multi unit housing)
Design Life	5,000 hours	50,000 hours
Cost	$50 - 100/ kW	$300 - 1000/ kW
Electrical Output	High Voltage AC or DC	48 V DC to 220 V AC
Efficiency	Very High	Very High
Power Range	20 to 1	10 to 1
Power Density (Volume)	Very High	Moderate
Specific Power (Weight)	Very High	Moderate
Operation	Intermittent	Continuous (24/7)
Energy Storage	Possibly	Possibly
Transient Response	1/10 of seconds	1/1000 of seconds
Short Term Fuel	Gasoline and diesel	Natural Gas/Propane
Long Term Fuel	Hydrogen	Hydrogen

Vision of hydrogen applications: Hydrogen will be available for every end-use energy need in the economy, including transportation, power generation, industrial process heaters, and portable power systems. Hydrogen will be the dominant fuel for government and commercial vehicle fleets. It will be used in a large share of personal vehicles and light duty trucks. It will be combusted directly and mixed with natural gas in turbines and reciprocating engines to generate electricity and thermal energy for homes, offices, and factories. It will be used in fuel cells for both mobile and stationary applications. And it will be used in portable devices such as computers, mobile phones, Internet hook-ups, and other electronic equipment.

Challenges

To achieve this vision for hydrogen applications, the following challenges will need to be overcome.

Transportation, Stationary, and Portable Applications Require Technological and Engineering Solutions

Transportation applications lack affordable and practical hydrogen storage with sufficient volumetric and gravimetric densities. The absence of a storage solution severely hinders investment in infrastructure development as different storage mediums could result in substantially different infrastructure strategies. There is also a lack of reliable, inexpensive, and efficient reformation technologies.

Customers Must Accept Hydrogen Technologies and Fuel Cell Vehicles

Fuel cell vehicles are in the early stages of development and the first vehicles are likely to fall short of consumer expectations (e.g., range, cold-weather capability). By comparison, conventional internal-combustion engine vehicles have had the benefit of more than 100 years of technological refinement as well as relatively reliable, low-cost gasoline to power them. Automakers continue to produce conventional cars that are progressively cleaner and more fuel-efficient.

The hydrogen fuel infrastructure for fuel cell vehicles (without on-board reformation) will be deployed gradually, and is likely to be severely limited in the early years. In the early stages of commercialization, automakers are also likely to put a limited offering of model types on the market, restricting consumer choices. Most of the benefits that hydrogen-fueled vehicles offer over conventionally fueled vehicles are societal in nature (e.g., reduced carbon dioxide and criteria emissions, energy security) and will not be fully realized until many years after market introduction. In addition, hybrid electric and gasoline vehicles will be coming on to the market in the near-term, and could compete with hydrogen-fueled vehicles.

Although conventional technologies (with modifications) can offer early market entry for hydrogen as an energy carrier, conventional technologies will continue to hold a major market advantage in terms of vehicle range. With their decreased energy density, today's hydrogen-fueled vehicles have ranges of 200 to 250 miles—far short of the 380 to 400 miles offered by contemporary hydrocarbon vehicles. Optimized hydrogen-fueled conventional technologies have been demonstrated that might achieve efficiencies close to the anticipated levels of fuel cells, with emissions that are near zero. Vehicles need to be redesigned for parity in range, or the public needs to be convinced of the benefits of a hydrogen-fueled vehicle to increase its perceived value.

> In transportation applications, reformer research should be directed to enable nearterm end use of hydrogen prior to the development of a nationwide hydrogen delivery system.

Paths Forward

Conduct Research and Development to Address Critical Challenges to a Hydrogen Vision

Transportation and stationary applications will require development of low-cost and durable fuel cell stacks and systems. Development needs include high-temperature membranes for fuel cells; low-cost, fast-response, and low-power-consumption sensors and controls; low-cost, reliable, subsystem components such as compressors, pumps, and power electronics; and low-cost, reliable, hybrid batteries and ultra-capacitors.

In transportation applications, reformer research should be directed to enable near-term end use of hydrogen prior to the development of a nationwide hydrogen delivery system. Hydrogen storage research for vehicles should focus on systems that have the capability to match the driving range of equivalent gasoline vehicles. Development should focus on systems that are safe, have low weight and small size, and are cost competitive. Storage systems will have to be compatible with the fueling infrastructure, and the safety of storage system designs should be ensured through the development of codes and standards.

Combustion strategies and after-treatments must be optimized to maximize power densities and thermal efficiencies while minimizing tailpipe emissions. Challenges in engineering design include developing flow handling and engine management systems for a commercial-ready device. Lean, premixed combustion is the preferred strategy to control emissions in stationary turbines. It allows for control over the combustion process, but frequently results in acoustic instabilities. Research is needed to develop better control strategies that will help hydrogen and hydrogen-enriched hydrocarbon fuels gain wider acceptance.

Conventional conversion devices need to be demonstrated in stationary, transportation, and hybrid stationary-mobile applications, and should be designed to promote the creation of hydrogen clusters.

For the near term, research is needed to address such issues as durability, cost of the fuel cell stack, system integration, system architecture, and reformer development. The Federal government should designate a lead fuel cell/hydrogen laboratory so that this work is concentrated and available widely. That entity should develop breakthrough technologies and fundamentals that are broadly applicable to all fuel cell applications, while also providing a facility for testing more robust products.

Increase Demonstrations Significantly

Demonstrations should showcase the near-term availability of multiple alternative technologies for distributed generation power parks. This effort could include the development of hydrogen-based mini economies around an existing hydrogen infrastructure.

Conventional conversion devices need to be demonstrated in stationary, transportation, and hybrid stationary-mobile applications, and should be designed to promote the creation of hydrogen clusters. As the number of demonstration projects grows, so will the hydrogen clusters. This will help jumpstart the creation of the hydrogen infrastructure for both stationary and transportation applications.

Relationships should be built and expanded beyond current demonstration activities. The Federal government could demonstrate stationary fuel cells in government buildings. Numerous partnerships, such as the U.S. Department of Energy's FreedomCAR program and the California Fuel Cell Partnership, have demonstrated the advantages of public-private cooperation for transportation applications. Public-private partnerships to demonstrate early vehicles and the associated fueling infrastructure will be necessary to minimize economic risks.

Institute Regulations, Codes and Standards to Foster Customer Acceptance of the Hydrogen Vision

Standard nationwide interconnection agreements are needed to enable connection to the current electrical grid without punitive costs, policies, or actions. Standard agreements and educational materials should be prepared for use by fire, insurance, and building code officials.

Develop Public Policies that Encourage Use of Hydrogen as a Fuel

Convincing Americans to use hydrogen applications will require incentives such as cost-sharing demonstrations, policies for price parity, and "rights-of-way" for hydrogen infrastructure (similar to those in the natural gas industry).

The Federal government should adopt national interconnection standards, require utilities to treat stationary hydrogen customers in a manner similar to others in the same rate class, and ensure that distributed generation options are valued for their ability to utilize waste heat and achieve high efficiencies. Strategies might include development of emissions trading that reaches the small size level, assigning value to externalities via a "carbon tax," or other such measures.

Government could also provide incentives for investing in new technologies, such as tax credits for transportation, stationary, and portable hydrogen systems and for hydrogen infrastructure development.

Hydrogen Energy Applications: Needs and Activities

Needs	Activities	Primary Entities
Consensus on technically-based codes and standards	Consortium of agencies to set codes and standards	Federal, State government agencies
Public-private partnership for systems demonstrations	Benchmark California programs demonstration of zero emission vehicle; fuel cell bus fleet demonstration	Federal, State, local government agencies
Government as early adopter	Begin demonstration programs	Federal agencies
Short-term hydrogen end-use technologies to stimulate infrastructure and market readiness	Generate the market with the purchase of fleets	Federal agencies
R&D for onboard storage	Include demos for onboard storage tanks that can achieve 6,7,8% weight of stored hydrogen per weight of tank	Federal agencies, National Laboratories, industry
Community-based clustered applications and installations	Issue RFP, select winners, commence building, complete installations	Federal grants; State and local government and industry cost-share

Conclusion

The ultimate aim is to enable consumers to use hydrogen energy devices for transportation, electric power generation in cities and homes, and portable power in electronic devices such as mobile phones and laptop computers. Once the cost and performance issues associated with hydrogen energy systems have been addressed, the next challenges will involve customer awareness and acceptance. Safety, convenience, affordability, and environmental friendliness are key consumer demands. Industry should focus its efforts on understanding consumer preferences and building them into hydrogen system designs and operations. Government (Federal and State) should identify opportunities to use hydrogen systems in facilities for distributed generation, combined heat and power, and vehicle fleets.

8. EDUCATION AND OUTREACH

Introduction

Educating consumers, industry leaders, and public policy makers about the benefits of hydrogen is critical to achieving the Vision. The development and implementation of broad-based advocacy and public education programs are critical components of success. Stakeholders lack any real understanding about the development and use of hydrogen: how is it used today; how can it be used in the future; what are the advantages of hydrogen in these new markets; how does hydrogen relate to renewable resources; and what are the storage and safety issues.

Public and private organizations can take a variety of approaches to increase nationwide awareness and acceptance of hydrogen, including coalition building, public relations and media campaigns, community demonstration projects, and long-term commitment of resources to America's education system. To insure success in this arena, consumers will need to understand the value and the relative risks of hydrogen; industry will need to work with other sectors and the media to create consumer demand for hydrogen; and public policy makers will need to develop consistent and sustainable policies and regulations that support hydrogen systems.

The challenge before us is to educate the Nation about the benefits of hydrogen, the relative costs and issues, and the path toward its eventual use as a ubiquitous energy carrier. The message that hydrogen is clean, secure, and safe must be effectively demonstrated to wide and varied audiences. Target audiences include consumers, students, educators, public policy makers, non-governmental organizations, the research and development community, industry, the media, multilateral institutions, and professional and trade associations.

Educating consumers, industry leaders, and public policy makers about the benefits of hydrogen is critical to achieving the Vision.

Challenges

A nationwide effort to promote hydrogen faces regulatory and institutional challenges. Common consumer misconceptions about hydrogen as a fuel can impede widespread acceptance of hydrogen. The following are identified as the most pressing issues to be addressed through outreach and education.

The Public Lacks Awareness

Consumers are generally unaware of hydrogen as an energy alternative. Since there is little consensus about the severity of today's environmental problems there is little impetus for change. Hydrogen needs to be "personalized" for consumers so that they understand the value of switching from fossil fuel-based energy systems to hydrogen systems.

Too Few Examples and Success Stories Exist to Lure Business Investment

Demonstrations of developing technologies are rare, leading to a lack of early adopters willing to invest in new technologies and commercialization. Without demonstrations, technologies remain unproven, consumer demand for new products remains weak, and financing and investment communities are slow to feed the capital pipeline.

Hydrogen Education Programs Are Minimal

A lack of structured education programs on hydrogen exists at all levels. Teacher training on the benefits of hydrogen has not been a priority, and students at all educational levels are not being introduced to hydrogen. As a result, students are not stimulated to pursue science and technology careers that support growing business interests, and do not share information about hydrogen and its benefits with their parents and peers.

Policies Are Inconsistent

Policy makers often are not knowledgeable about hydrogen as a fuel, nor do they understand how it works. Inconsistent regulations at the Federal, state, and local levels—including inconsistent or nonexistent codes and standards—are barriers to the widespread implementation of hydrogen. In addition, existing policies fail to accurately measure the true costs and environmental impacts of our energy choices, thus sending incomplete messages to consumers.

Target Outreach Audiences

- Consumers (residential, commercial, industrial)
- Students and educators (K-12 and college students, science teachers)
- Federal, state, and local officials/policymakers (including transit agencies, code officials, public utility commissions, state energy offices, and regional planning organizations)
- Non-governmental organizations, foundations, environmental groups, and institutional interests (lenders, investors, insurance, real estate)
- The R andD community
- Industry groups (including industry executives, service station operators and owners, and vehicle fleet operators/owners)
- Multilateral institutions (World Bank, development banks, etc.)
- Professional and trade associations

Consumers Harbor Safety Concerns

Consumers may unnecessarily fear hydrogen if they are misinformed about its safety, and may hold misconceptions about the risk of using it in homes, businesses, and automobiles. Fear may also stem from a lack of understanding about the dangers associated with fuels that consumers use today. The following message needs to be communicated: like all fuels, hydrogen can be handled and used safely with appropriate sensing, handling, and engineering measures.

Paths Forward

Specific actions must be taken to overcome the barriers and achieve the vision for a hydrogen economy.

Establish Regional, State, and Local Networks

Networks should be developed to include code officials, building engineers, energy regulators, and consumers in hydrogen technology demonstrations. These networks should provide public education on installation, codes and standards, and safety issues.

Create a Broad Coalition to Influence U.S. Energy Policy on Hydrogen

A hydrogen advocacy coalition could be created to support public policies that encourage the development of hydrogen production, storage, and utilization technologies; the removal of key regulatory and market barriers; the development of education curricula; and the creation of public policies that would make hydrogen an important component of a secure, efficient, and environmentally acceptable energy mix. The coalition could reach out to public and private decision makers regarding the need to implement consistent and sustainable policies and procedures that support hydrogen systems. It could also encourage regional hydrogen initiatives and partnerships, establish informational caucuses, and support a continuous path of technological improvement.

Develop a Comprehensive Public Education and Outreach Program

Hydrogen needs to get "on the map" and in the minds of consumers. Getting the message out will require a coordinated effort by government, industry, and non-profit communities to develop a broad-based education and outreach program. This program, which should be developed as soon as possible, should include public relations and advertising campaigns. Public spokespersons need to be identified and media briefing packets produced. A product recognition tag, similar to EnergyStar®, should be developed, and hydrogen success stories should be touted. Other public relations and outreach activities would include:

- Construction of traveling exhibits on hydrogen
- Expansion of online hydrogen databases and information center
- Creation of compact disks and Internet marketing materials

Key components of the education and public relations program include the creation of effective consumer messages, awareness campaigns, and media outreach. Innovative ideas and creative incentives are needed to prime the population for migration toward a hydrogen economy. Consumers need to feel compelled to learn more about hydrogen and must be clear about how a hydrogen economy can benefit the environment and energy security of the Nation.

Hydrogen needs to be "branded" and "personalized" for the consumer; safety needs to be stressed. Messages need to be consistent (e.g., "Hydrogen is the Freedom Fuel," "Hydrogen—It Works," or "Hydrogen is 'The Power'").

In addition, industry should work with filmmakers to include product placement in movies. Community models and exhibits should also be developed to promote consumer participation and action.

Key components of the education and public relations program include the creation of effective consumer messages, awareness campaigns, and media outreach.

Create a Public Demonstration Hydrogen Village

Homebuilders, architects, lending institutions, realtors, technology manufacturers, and related associations should lead an effort to launch a community model or hydrogen village that identifies stakeholders, products, and the infrastructure of a hydrogen economy. Multiple villages, in whole or in part, could be situated in strategic locations across the United States as instructional models for outreach programs directed toward students, government, and industry.

Commit Resources for Long-Term Education of Students at All Levels

Student education is a key component to broadcasting the hydrogen message and developing a knowledgeable, involved hydrogen support network. Without a targeted technology (and applications-level) education program for students and teachers, our past will continue to define our future. Long-term resources should be committed to educate all students. Easy-to-integrate curricula should be developed for kindergarten to grade 12, vocational, four-year engineering, and advanced-degree students. Hydrogen education packages should be created, including lesson plans, videos, demonstration hardware, and experiments to help educate science teachers and their students.

Educator training should be made available to all interested teachers through summer workshops and in-service training. Prizes could be offered for college-level engineering theses and projects on vehicle systems, stationary applications, and storage technologies. In addition, a hydrogen fellowship program should be created to encourage interest in the industry at the graduate-level. Lead organizations for this effort include the National Science Teachers Association, the U.S. Department of Education, education agencies and boards, and textbook publishing companies. This effort should begin immediately with an inventory of educational resources and development of teacher training materials that can be integrated with existing energy education materials.

Hydrogen Themes

• Hydrogen is "The Freedom Fuel"
• Hydrogen provides independence and an environmental choice
• Hydrogen solves foreign oil dependency and improves the environment
• Hydrogen is everywhere—"it's right in our backyard"
• A hydrogen economy includes other fuels
• Hydrogen—it works (it is an ongoing business today)
• Hydrogen is safe
• Hydrogen is a long-term energy solution
• Hydrogen is the "man on the moon" equivalent for this generation

Conclusion

Education and outreach on the many benefits of hydrogen is a vital element of this Roadmap. It will require a long-term, coordinated commitment by diverse stakeholders to effectively communicate key hydrogen messages to a wide and varied audience. A broad-based education and outreach program—including public relations, media campaigns, demonstration activities, and policy initiatives—must start immediately. Education is an ongoing process impacting all aspects of the hydrogen roadmap and its prospects for success.

9. CONCLUSIONS

The fundamental purpose of this Roadmap is to define a common set of objectives and activities agreed upon by government, industry, universities, National Laboratories, environmental organizations, and other interested parties. Focusing resources on this common agenda will facilitate evaluation of a hydrogen economy and potentially stimulate investment in the development of a hydrogen energy system.

Strong government-industry partnerships are needed to evaluate the potential for hydrogen to play a larger role in America's energy future. This undertaking requires government leadership and a significant, long-term investment of the Nation's resources, both public and private.

Adoption of hydrogen can be encouraged by policies that reflect the external costs of energy supply, security, air quality, and global climate. These policies must be consistent and provide a clear signal to industry and consumers. A societal dialogue should be initiated to stimulate an informed, ongoing discussion of how we as a society value a low-polluting and diverse energy supply. Although the United States is the focus of this Roadmap, energy markets are global. Over the next several decades, much of the global growth in energy demand is projected to be in developing countries. U.S. efforts in hydrogen technologies can have global benefits.

The widespread use of hydrogen will impact every aspect of the U.S. energy system, from production through end-use. The individual components of a hydrogen energy system— production, delivery, storage, conversion and end-use applications—are closely interrelated and interdependent. The design and implementation of a hydrogen economy must be considered at the "whole system" level.

> The fundamental purpose of this Roadmap is to define a common set of objectives and activities agreed upon by government, industry, universities, National Laboratories, environmental organizations, and other interested parties.

Elements of a Hydrogen Energy System

Production—Government-industry coordination on hydrogen production systems is required to lower overall costs, improve efficiency, and reduce the cost of carbon sequestration. Better techniques are needed for both central-station and distributed hydrogen production. Efforts should focus on improving existing commercial processes such as steam methane reformation, multifuel gasification, and electrolysis. Development should continue on advanced production techniques such as biological methods and nuclear- or solar-powered thermochemical water-splitting.

Delivery—A greatly expanded distributed infrastructure will be needed to support the expected development of hydrogen production, storage, conversion, and applications. Initial efforts should focus on the development of better components for existing delivery systems, such as hydrogen sensors, pipeline materials, compressors, and high-pressure breakaway hoses. Cost, safety, and reliability issues will influence the planning, design, and development of central versus distributed production and delivery. To address the "chicken and egg" (demand/supply) dilemma, demonstrations should test various hydrogen infrastructure components for both central and distributed systems in concert with end-use applications (e.g., fueling stations and power parks).

Storage—Hydrogen storage is a key enabling technology. None of the current technologies satisfy all of the hydrogen storage attributes sought by manufacturers and end users. Government-industry coordination on research and development is needed to lower costs, improve performance, and develop advanced materials. Efforts should focus on improving existing commercial technologies, including compressed hydrogen gas and liquid hydrogen, and exploring higher-risk storage technologies involving advanced materials (such as lightweight metal hydrides and carbon nanotubes).

Conversion—Conversion of hydrogen into useful forms of electric and thermal energy involves use of fuel cells, reciprocating engines, turbines, and process heaters. Research and development are needed to enhance the manufacturing capabilities and lower the cost of fuel cells as well as to develop higher-efficiency, lower-cost reciprocating engines and turbines. Efforts should focus on developing profitable business models for distributed power systems, optimizing fuel cell designs for mobile and stationary applications, and expanding tests of hydrogen-natural gas blending for combustion. Research is required to expand fundamental understanding of advanced materials, electrochemistry, and fuel cell stack interfaces and to explore the fundamental properties of hydrogen combustion.

Applications—Ultimately, consumers should be able to use hydrogen energy for transportation, electric power generation, and portable electronic devices such as mobile phones and laptop computers. Cost and performance issues associated with hydrogen energy systems will need to be addressed in tandem with customer awareness and acceptance. Key consumer demands include safety, convenience, affordability, and environmental friendliness. Efforts should focus on understanding consumer preferences and building them into hydrogen system designs and operations. Opportunities should be identified to use hydrogen systems in facilities for distributed generation, combined heat and power, and vehicle fleets. Supportive energy and environmental policies should be implemented at the Federal, State, and local levels.

Education and outreach—Hydrogen energy development is a complex topic, and people are uncertain about impacts on the environment, public health, safety, and energy security. Ultimately, consumer preferences drive the choices made in energy markets, technology development, and public policy. Informing the public through educational and training materials, science curricula, and public outreach programs will help garner public acceptance for hydrogen-related products and services.

Codes and standards—Uniform codes and standards for the design, manufacture, and operation of hydrogen energy systems, products, and services can dramatically speed the development process from the laboratory to the marketplace. Government-industry coordination can accelerate codes and standards processes, which must also span national boundaries and be accepted by international bodies to achieve global acceptance.

Final Thoughts

Development of hydrogen energy technologies represents a potential long-term energy solution with enormous benefits for America. A coordinated and focused effort is necessary to bring public and private resources to bear on evaluating the costs and benefits of the transition to a hydrogen economy. Next steps will include the development of detailed research and development plans for each of the technology areas. A significant commitment of resources—funding, people, and facilities—will be needed to accomplish this. Specifically:

Science and Engineering

Challenges are presented in the fundamentals of the materials sciences, electrochemistry, biology, engineering design, and manufacturing. The Federal government, with assistance from State energy agencies and researchers in universities and National Laboratories, has a critical role to play in advancing scientific frontiers. Breakthroughs in hydrogen production, storage, and conversion could alter the assessment of costs and benefits substantially. Greater emphasis needs to be placed on focusing resources on the most promising opportunities. Unproductive research pathways should be terminated. Dissemination of results and the transfer of knowledge to the private sector need to occur at a faster pace and in a more effective manner.

Technology Development

An increased level of technical coordination among industry, government, universities, and the National Laboratories is needed to advance hydrogen-related technology development programs in accordance with the activities and priorities put forth in this Roadmap. More specific technology roadmaps need to be developed to attract government and industrial funding to pursue specific technology opportunities, particularly in the areas of hydrogen production, storage, and conversion. The Federal government needs to more closely coordinate activities across several agencies including the Departments of Energy, Transportation, Commerce, and Defense.

Demonstrations

Immediate government-industry coordination is needed in order to implement several types of hydrogen energy technologies — spanning stationary, mobile, and portable applications—to evaluate the potential of hydrogen as an energy solution for America. Technology demonstrations and hydrogen pilot projects can help to uncover problems and compile empirical data to better estimate the costs and benefits of infrastructure requirements for transition to a hydrogen economy. Existing efforts to implement hydrogen fueling stations, power parks, and municipal fleets, for example, should be replicated in a variety of locations and climate conditions. Results should be properly documented and disseminated widely.

Institution Building

Energy and environmental policies, utility regulations, business practices, and codes and standards are critical elements of the institutional infrastructure needed to develop hydrogen energy. Public outreach and education programs could inform the public on the ways existing institutions support hydrogen energy development. Policy analysis is needed to identify unnecessary regulatory barriers, and business analysis is needed to identify profitable models for hydrogen energy development (and related concepts such as distributed energy generation and combined heat and power). Efforts to engage code officials in the United States and throughout the world should increase to foster greater harmony and consistency with regard to hydrogen energy products and services.

Implementation of this Roadmap involves making progress on the top priority actions and recommendations. Only by working together—government, industry, universities, National Laboratories, and environmental organizations—will progress be made.

> Energy and environmental policies, utility regulations, business practices, and codes and standards are critical elements of the institutional infrastructure in which hydrogen energy needs to develop.

A. Vision and Roadmap Participants

Activated Metals Technologies,
Richard Uchrin

Air Liquide America, Gary McDow

Air Products and Chemicals, Inc.,
Arthur Katsaros, Robert Miller, Venki Raman, Chris Sutton

Alameda Contra Costa Transit,
Jaimie Levin

Alberta Energy Research Institute,
Surindar Singh

American Petroleum Institute,
Ron Chittim, David Lax, Prentiss Searles, James Williams

Antares Group, Inc., David Gelman
Argonne National Laboratory, Balu Balachandran, Richard Doctor,
George Fenske, James Miller, Marianne Mintz

ATK Thiokol Propulsion, Andrew Haaland

ATTIK, Ferris Kawar

Avista Labs, Inc., J. Michael Davis

Ballard Power Systems, Stephen Kukucha

Battelle-Pacific Northwest National Laboratory, Jae Edmonds

BOC Group, The, Satish Tamhankar

BP, Lauren Segal

California Air Resources Board, Shannon Baxter

California Energy Commission, Susan Brown, Louise Dunlap, Kenneth
Koyama

ChevronTexaco Ovonic Hydrogen Systems, LLC, Rosa Young

ChevronTexaco Technology Venture, Robert Dempsey, Gene Nemanich, Greg Vesey

Clean Power Engineering Company, David Bruderly

Consultant, James Hempstead

Covington and Burling, Kipp Coddington

CryoFuel Systems, Inc., John Barclay

DaimlerChrysler, William Craven, Stephen Zimmer

DCH Technology, Inc., John Donohue, Mary Rose Szoka de Valladares

Deere and Company, Jason Francque, Bruce Wood

Denali Commission, Jeffrey Staser

DuPont Company, Patricia Watson

Dykema Gossett, PLLC, Richard Bradshaw

Dynetek Industries, Ltd., Robb Thompson

E2I, Adriene Wright

Electric Power Research Institute, Neville Holt, Revis James

Energetics, Incorporated, Jeannette Brinch, Ross Brindle, Tracy Carole, Jack Eisenhauer, Lauren Giles, Jamie McDonald, Jennifer Miller, Tara Nielson, Rich Scheer, Edward Skolnik, Charles Smith, Christina TerMaath

Energy Conversion Devices, Krishna Sapru

Entergy Nuclear, Inc., Dan Keuter, Stephen Melancon

Ergenics, Inc., David DaCosta

Eversole Associates, Kellye Eversole

ExxonMobil Refining and Supply Company, William Lewis

Fauske and Associates, Inc., Wison Luangdilok

Florida Hydrogen Business Council, Marshall Gilmore

Ford Motor Company, Frank Balog, Stephen Fan

Fuel Cells 2000, Maria Bellos

Gas Equipment Engineering Company, Martin Shimko

Gas Technology Institute, Patrick Findle, Francis Lau, Michael Romanco

General Atomics, Gottfried Besenbruch, Ken Schultz

General Electric, Sanjay Correa, Ravi Kumar, Daniel Smith

General Motors, Byron McCormick, Kenneth Cameron, Jerry Rogers

Global Environment and Technology Foundation, F. Henry Habicht

H2 Solutions, Inc., E.J. Belliveau, Ed Bless

H2Gen Innovations, Inc., C.E. Thomas

Hamilton Sustrand, Michael Gan, Tedd Lima

Hart Downstream Energy Services, Stacy Klein

Hawaii Natural Energy Institute, Richard Rocheleau

History Associates, Inc., Rodney Carlisle, James Lide

Honda RandD Americas, Inc., Ben Knight, Shiro Matsuo

House Committee on Science, John Darnell

Houston Advanced Research Center, George King

Hydrogen 2000, Inc., Susan Leach

Hydrogen and Fuel Cell Letter, The, Peter Hoffmann

Hydrogen Now!, Katie Hoffner

Hydrogen Technical Advisory Panel, Helena Chum, Michael Hainsselin, Chung Liu, David Nahmias

IET, Adam Cohn

Innovative Design, Inc., Michael Nicklas

Inside Washington Publishers, Peter Rohde

Institute for Alternative Futures, Robert Olson

JMC, Inc., John Baker, Kenichi Sakamachi

Johnson Matthey Fuel Cells, Wilson Chu

K. Wetzel and Company, Kyle Wetzel

K. Winn and Associates, Kathleen Winn

Lincoln Composites, Brent Gerdes, Richard Rashilla, Dale Tiller

Los Alamos National Laboratory, Kenneth Stroh
Marathon Ashland Petroleum, LLC, Michael Leister, Faruq Marikar

Massachusetts Institute of Technology, Daniel Cohn

Massachusetts Renewable Energy Trust, Raphael Herz

Methanol Institute, Greg Dolan

Millennium Cell, Katherine McHale, Christine Messina-Boyer, Stephen Tang

Minnesota Corn Growers Association, Duane Adams

NASA Kennedy Space Center, David Bartine

National Academy of Sciences, Peter Blair, Brendan Dooher, James Zucchetto

National Energy Technology Laboratory, Rita Bajura, Anthony Cugini, Richard Noceti, F. Dexter Sutterfield

National Governors Association, Ethan Brown

National Hydrogen Association, Karen Miller, Jeffrey Serfass

National Institute of Standards and Technology, Adam Pivovar, Terrence Udovic

National Renewable Energy Laboratory, Richard Truly, Susan Hock, Margaret Mann, Jim Ohi, Catherine Gregoire-Padro, Terry Penney, John Turner

Natural Resources Canada, Vesna Scepanovic

Natural Resources Defense Council, Daniel Lashof

NiSource Inc., Arthur Smith

Northeast Midwest Institute, Suzanne Watson

Oak Ridge National Laboratory, Tim Armstrong, Robert Hawsey

Office of Assistant Secretary of the Navy (IandE), Leo Grassilli

Office of Naval Research, Richard Carlin

Office of Science and Technology Policy, Gene Whitney

Office of Senator Akaka, Jaffer Mohiuddin

Pacific Carbon International, Kay Fabian

PDVSA-Citgo, Manuel Pacheco

Pinnacle West Capital Corporation, Peter Johnston

Plug Power, William Ernst, Jennifer Schafer

Praxair, Inc., Donald Terry, Ed Danieli

Princeton University, Joan Ogden

Proton Energy Systems, Inc., Robert Friedland, William Smith

Quantum Technologies, Alan Niedzwiecki, Andris Abele

QuestAir Technologies, Inc., Mark Grist, Edson Ng

SAIC, Richard Sassoon

Sandia National Laboratories, Donald Hardesty, Jay Keller, George Thomas

Savannah River Technology Center, William Summers

Sentech, Inc., Erin Cready, Jonathan Hurwitch, George Kervitsky, Rajat Sen

Shell Hydrogen, Brad Smith

South Coast Air Quality Management District, Norma Glover

Southern Hydrogen Fuel Cell Research Partnership, Andrew Searcy

Startech Environmental Corporation, John Celentano, Joseph Klimek

Structural Composites Industries, Randolph Schaffer

Stuart Energy Systems, Matthew Fairlie, Paul Scott, Andrew T.B. Stuart

SunLine Transit Agency, Richard Cromwell, William Clapper

Technology and Management Services, Inc., Mark Ackiewicz

Teledyne Energy Systems, Inc., Jay Laskin

Texaco Energy Systems, Graham Batcheler

Underwriters Laboratories, Gordon Gillerman

U.S. Department of Energy, Christopher Bordeaux, Robert Card, Lucito Cataquiz, Steven Chalk, James Daley, Robert Dixon, Kathi Epping, Nancy Garland, David Garman, Thomas Grahame, Sigmund Gronich, Tom Gross, Art Hartstein, David Henderson, Donna Ho, Arun Jhaveri, R. Shane Johnson, Matthew Kauffman, Karen Kimball, Kyle McSlarrow, C. Lowell Miller, JoAnn Milliken, Richard Moorer, Zdenek Nikodem, William Parks, Samuel Rosenbloom, Neil Rossmeissl, Edward Schmetz, Frank (Tex) Wilkins

U.S. Environmental Protection Agency, John Wysor

U.S. House of Representatives, Congressman Roscoe Bartlett

U.S. Hydrogen, Charles Veley

U.S. TAG ISO TC 197, Robert Mauro

University of Central Florida, Gregory Schuckman

University of Illinois, Alexander Fridman

University of Michigan, Arvind Atreya

University of Nevada-Reno, Dhanesh Chandra

University of South Carolina, James Ritter

UTC Fuel Cells, William Miller, Doug Wheeler

Verizon, Thomas Bean

Westinghouse Savannah River Company, Melvin Buckner

Wexler and Walker Public Policy Associates, Robert Walker, Peter Holran, Jodi Salup

World and I Magazine, Glenn Strait

World Resources Institute, James MacKenzie

Worldwatch Institute, Seth Dunn

ZECA Corporation, Alan Johnson

Ztek Corporation, David Tsay

B. Acronyms

AC	Alternating Current
ANSI	American National Standards Institute
ASME	American Society of Mechanical Engineers
ATS	Advanced Turbine Systems
CHP	Combined Heat and Power
CSA	Canadian Standards Association
DC	Direct Current
DOD	U.S. Department of Defense
DOE	U.S. Department of Energy

DOT	U.S. Department of Transportation
EPA	U.S. Environmental Protection Agency
FY	Fiscal Year
ICC	International Codes Council
IEC	Independent Electrical Contractors
IEEE	Institute of Electrical and Electronic Engineers
ISO	International Standards Organization
kW	Kilowatt
NASA	National Aeronautics and Space Administration
PEM	Polymer Electrolyte Membrane
QC	Quality Control
RandD	Research and Development
RFP	Request for Proposal
SAE	Society of Automotive Engineers
SMR	Steam Methane Reformer
SOFC	Solid Oxide Fuel Cell
UL	Underwriters Laboratories, Inc.
V	Volt

In: Hydrogen Energy
Editor: A.O. Backus, pp. 75-111

ISBN 1-59454-733-5
© 2006 Nova Science Publishers, Inc.

Chapter 3

A NATIONAL VISION OF AMERICA'S TRANSITION TO A HYDROGEN ECONOMY —TO 2030 AND BEYOND

United States Department of Energy

A CALL FOR PARTNERSHIP

This document outlines a vision for America's energy future—a more secure nation powered by clean, abundant hydrogen. This vision can be realized if the Nation works together to fully understand hydrogen's potential, to develop and deploy hydrogen technologies, and to produce and deliver hydrogen energy in an affordable, safe, and convenient manner.

President Bush's National Energy Policy says, "In the long run, alternative energy technologies such as hydrogen show great promise." In response, Energy Secretary Spencer Abraham recently stated, "The President's Plan directs us to explore the possibility of a hydrogen economy." This vision document is a first step. It will be used as a foundation for formulating the elements of a National Hydrogen Energy Roadmap.

Hydrogen is a long-term solution to America's energy needs, with near-term possibilities. Efforts to achieve our energy goals need to begin now and continue with a sustained commitment over the next several decades. It is important to strengthen our partnerships among government, industry, and others in order to improve hydrogen technologies and systems, and to build the infrastructure—both physical and institutional—that will be needed in the years ahead. The tragic events of September 11, 2001, highlight the urgency of making progress toward greater energy security.

The efforts of the 53 business executives, Federal and State energy policy officials, and leaders of universities, environmental organizations, and National Laboratories who contributed to the development of this vision document by offering their views at the National Hydrogen Vision Meeting are deeply appreciated. A nationwide effort to achieve the hydrogen vision can only succeed through strong public-private partnerships, to address the issues involved in the introduction of a new vehicle infrastructure and distributed generation systems. It is a broad and encompassing task with major national benefits. All affected stakeholders need to join together in this most important endeavor.

EXECUTIVE SUMMARY

On November 15-16, 2001, 53 senior executives representing energy and transportation industries, universities, environmental organizations, Federal and State government agencies, and National Laboratories met to discuss the potential role for hydrogen systems in America's energy future. (A list of the participants can be found in the appendix.) The intent of the meeting was to identify a common vision of the "hydrogen economy," the time frame in which such a vision could be expected to occur, and the key milestones that would need to be accomplished to get there.

Based on the ideas and suggestions put forth by the participants during the meeting, this document presents a national vision for hydrogen to become a premier energy carrier, like electricity, for Americans. It will be used by various stakeholders including industry, policy makers, and researchers as the National Vision coordinating foundation for formulating future actions leading to a hydrogen economy. The meeting proceedings, which include the presentations and summaries of the notes from the facilitated breakout sessions, can be downloaded at www.eren.doe.gov/hydrogen.

Major Findings

- Hydrogen has the potential to solve two major energy challenges that confront America today: reducing dependence on petroleum imports and reducing pollution and greenhouse gas emissions.
- There is general agreement that hydrogen could play an increasingly important role in America's energy future. Hydrogen is an energy carrier that provides a future solution for America. The complete transition to a hydrogen economy could take several decades.
- The transition toward a so-called "hydrogen economy" has already begun. We have a hydrocarbon economy, but we lack the know-how to produce hydrogen from hydrocarbons and water, and deliver it to consumers in a clean, affordable, safe, and convenient manner as an automotive fuel or for power generation.
- The "technology readiness" of hydrogen energy systems needs to be accelerated, particularly in addressing the lack of efficient, affordable production processes; lightweight, small volume, and affordable storage devices; and cost-competitive fuel cells.
- There is a "chicken-and-egg" issue regarding the development of a hydrogen energy infrastructure. Even when hydrogen utilization devices are ready for broad market applications, if consumers do not have convenient access to hydrogen as they have with gasoline, electricity, or natural gas today, then the public will not accept hydrogen as "America's clean energy choice."

CONCLUSIONS

- To achieve significant progress toward the development of hydrogen energy systems, Federal and State governments need to implement and sustain consistent energy policy actions that elevate hydrogen as a priority to meet energy security, energy independence, and climate change challenges.
- A successful effort to achieve the hydrogen vision will require a strong public-private partnership on hydrogen energy development, as called for in the President's National Energy Policy. Efforts must be focused on finding new ways to collaborate on the development and use of hydrogen energy. For example, the Federal government could play a valuable role as first-use "customer."

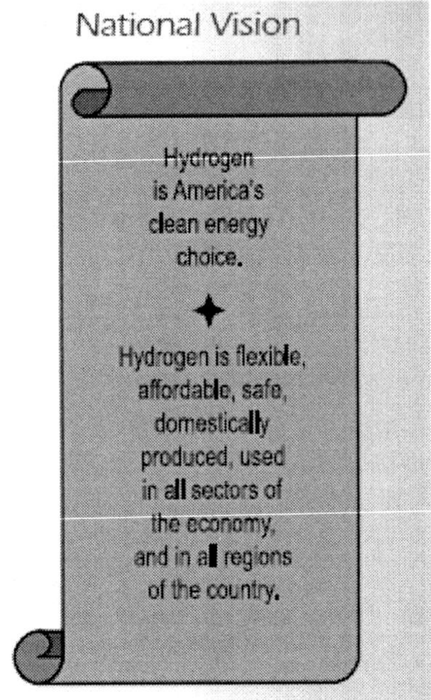

Overview of the Transition to the Hydrogen Economy

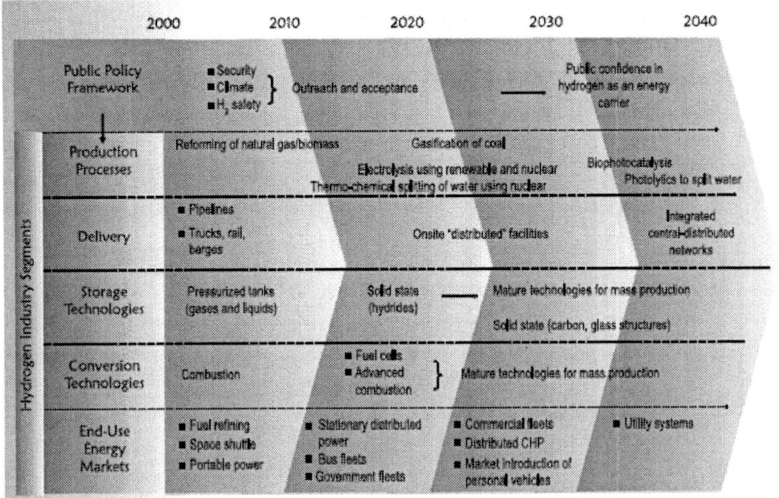

- . A logical next step is the collaborative development of a National Hydrogen Energy Roadmap. This process needs to address research, development, testing, public outreach and education, and codes and standards for hydrogen production, delivery, and use. It should cover both mobile and stationary applications of hydrogen energy systems, and it should involve industry, government, universities, and the National Laboratories.

1. INTRODUCTION

More than fifty executives from energy-related and transportation industries, Federal and State government agencies, universities, environmental organizations, and National Laboratories participated in the "National Hydrogen Vision Meeting," which was held in Washington, DC, on November 15 and 16, 2001. (A list of the participants can be found in the appendix.) This document reflects the ideas and priorities put forth by the meeting participants.

The meeting was held in response to specific recommendations in the Bush Administration's *National Energy Policy*, which was released on May 17, 2001. This comprehensive energy strategy contains 105 recommendations for securing America's energy future, including the expansion of energy supplies, improvement of infrastructure, modernization of energy conservation, and protection of the environment. In considering long-term energy and climate change solutions, hydrogen is singled out as a "future energy source...that shows great promise...and is compatible with existing energy technologies, such as fuel cells, engines, and combustion turbines." The report recommends that the U.S. Department of Energy "focus research and development efforts on integrating current programs regarding hydrogen, fuel cells, and distributed energy."

A National Hydrogen Vision and Roadmap process was initiated by the U.S. Department of Energy, in partnership with key hydrogen energy and transportation organizations, to address this recommendation. The goals of the vision and roadmap are to identify areas of agreement and disagreement about hydrogen's role in America's energy future, to establish a timeframe in which a hydrogen economy may be realized, and to discuss alternative pathways to achieve the vision. Special emphasis has been placed on public-private partnerships, and joint research, development, demonstration, education, and outreach programs.

While there have been many successful technical meetings on hydrogen research and development programs over the past decade, including semi-annual meetings of the Hydrogen Technical Advisory Panel, this meeting marked the first occasion that a broad cross-section of senior business leaders and energy and environmental officials from across the United States have met to discuss hydrogen energy development and its future as an energy source for America.

A broad range of stakeholders were invited to participate in the vision meeting. The participating non-Federal organizations included energy companies, automobile companies, fuel cell manufacturers, hydrogen equipment manufacturers, State agencies, and environmental organizations. Participating Federal organizations included the Department of Energy, the U.S. Navy, and the National Aeronautics and Space Administration.

The proceedings of the vision meeting, which includes copies of presentations and summaries of the notes from the discussions, is available at www.eren.doe.gov/hydrogen.

The meeting included discussions about the following specific topics:

- Status of today's U.S. hydrogen energy industry
- Factors—both supporting and inhibiting—that will shape future hydrogen energy development
- A vision for the future of the hydrogen energy industry

- What is meant by the term "hydrogen economy," and the most likely time frame(s) in which it may take place
- The key milestones and pitfalls that are likely to be encountered on the path to a "hydrogen economy"
- Ways in which industry and government can form stronger partnerships to address hydrogen energy development

The vision and roadmap process has been a useful technique for organizing research and development partnerships involving the U.S. Department of Energy, industry, the National Laboratories, and universities. The process typically involves a series of meetings attended by industry leaders and visionaries, technical experts, and practitioners to discuss future needs using professional facilitators to guide the discussions. Executive-level participation was emphasized in the development of the vision, and technical managers and practitioners will be encouraged to participate in the development of the roadmap.

2. THE HYDROGEN INDUSTRY TODAY

The current hydrogen industry is not focused on the production or use of hydrogen as an energy carrier or a fuel for energy generation. Rather, the nine million tons of hydrogen produced each year are used mainly for chemicals, petroleum refining, metals, and electronics. For example, the processes for making gasoline and diesel fuels, such as the breakdown of heavier crude oils and the removal of sulfur, are major users of hydrogen. The production of ammonia, used to make fertilizers, also consumes large amounts of hydrogen.

The use of hydrogen as an energy carrier or major fuel requires development in several industry segments, including production, delivery, storage, conversion, and end-use. The table below provides a list of terms and explanations for hydrogen energy systems. Each industry segment is integral to building a hydrogen-based economy, and the development of one segment relies on corresponding development of all other segments.

Elements of Today's Hydrogen Energy System

Hydrogen Industry Segment	Explanation
Production	• The production of hydrogen from fossil fuels, biomass, or water • Involves thermal, electrolytic, and photolytic processes
Delivery	• The distribution of hydrogen from production and storage sites • Involves pipelines, trucks, barges, and fueling stations
Storage	• The confinement of hydrogen for delivery, conversion, and use • Involves tanks for both gases and liquids at ambient and high pressures • Involves reversible and irreversible metal hydride systems
Conversion	• The making of electricity and/or thermal energy • Involves combustion turbines, reciprocating engines, and fuel cells
End-Use Energy Applications	• The use of hydrogen for portable power in devices such as mobile phones and computers • The use of hydrogen for transportation systems such as fuel additives, fuel-cell vehicles, internal combustion engines, and in propulsion systems for the space shuttle • The use of hydrogen for stationary energy generation systems, including mission critical, emergency, and combined heat and power applications

Overview

Hydrogen can be produced through thermal, electrolytic, or photolytic processes applied to fossil fuels, biomass, or water. Renewable and nuclear systems can produce hydrogen from water using thermal or electrolytic processes. The thermal production process, which uses steam to produce hydrogen from natural gas or other light hydrocarbons, is most common. This hydrogen is either consumed on site ("captive" hydrogen) or distributed via pipelines or trucks ("merchant" hydrogen). Hydrogen can be stored in its elemental form as a liquid, gas, or as a chemical compound, and is converted into energy through fuel cells or by combustion in turbines and engines.

Each of these components of the hydrogen industry is under development. The following sections explain the current status of these technological areas in greater detail.

Production

Although hydrogen is the most abundant element in the universe, it does not naturally exist in its elemental form on Earth. It must be produced from other compounds such as water, biomass, or fossil fuels. Each method of production from these constituents requires energy in some form, such as heat, light, or electricity, to initiate the process.

In the United States, approximately 95 percent of hydrogen is currently produced via steam reforming.[1] Steam reforming is a thermal process, typically carried out over a nickel-based catalyst, that involves reacting natural gas or other light hydrocarbons with steam. This is a three-step process that results in a mixture of hydrogen and carbon dioxide, which is then separated by pressure swing adsorption, to produce pure hydrogen. Steam reforming is the most energy efficient commercialized technology currently available, and is most cost-effective when applied to large, constant loads. Research is being conducted on improving catalyst life and heat integration, which would lower the temperatures needed for the reformer and make the process even more efficient and economical.

Partial oxidation (autothermal production) of fossil fuels is another method of thermal production. It involves the reaction of fuel with a limited supply of oxygen to produce a hydrogen mixture, which is then purified. Partial oxidation can be applied to a wide range of hydrocarbon feedstocks, including light hydrocarbons as well as heavy oils and hydrocarbon solids. However, it has a higher capital cost because it requires pure oxygen to minimize the amount of gas that must later be treated. In order to make partial oxidation cost effective for the specialty chemicals market, lower cost fossil fuels must be used. Current research is aimed at improving membranes for better separation and conversion processes in order to increase efficiency, and thus decrease the consumption of fossil fuels.

Hydrogen can also be produced by using renewable and nuclear resources to extract hydrogen from water, but these methods are currently not as efficient or cost effective as using fossil fuels. Biomass can be thermally processed through gasification or pyrolysis to produce hydrogen.

---

[1] Air Products and Chemicals, Inc.

Research on nuclear-based hydrogen production is mostly conducted on thermo-chemical processes, which makes use of high reactor exit temperatures. Both are continuing to be developed. Creation of more efficient, less expensive electrolyzers using renewables and nuclear power is also ongoing.

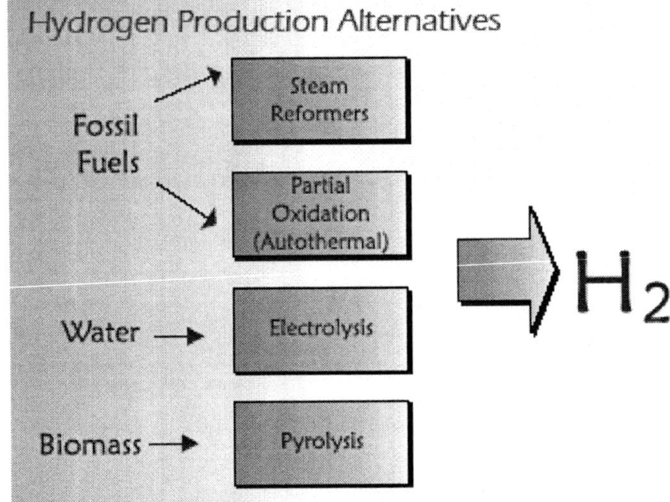

Delivery

A hydrogen energy infrastructure would include production and storage facilities, structures and methods for transporting hydrogen, fueling stations for hydrogen-powered applications, and technologies that convert the fuel into energy through end-use systems that power buildings, vehicles, and portable applications. This section focuses on existing infrastructure that moves the hydrogen from its point of production to an end-use device.

Today hydrogen is produced primarily in decentralized locations and is used on-site for making chemicals or upgrading fuels. Approximately 17 percent of hydrogen is centrally produced for sale and distribution, and is transported through pipelines or via cylinders and tube trailers.[2] Air Products and Chemicals Inc., Air Liquide Group, Praxair Inc., and the BOC Group are major producers of merchant hydrogen. Together these companies operate

---

[2] Air Products and Chemicals, Inc.

about 80 plants in the United States that are dedicated to the production of merchant hydrogen.

Similar to natural gas distribution, pipelines are used to supply hydrogen to customers. Currently hydrogen pipelines are used in only a few areas of the United States. Air Liquide Group, Air Products and Chemicals Inc., and Praxair Inc. operate hydrogen pipelines in Texas, Louisiana, California, and Indiana. Pipelines provide an efficient means for transporting hydrogen. Concerns regarding the weakening of carbon steel pipes in a process called hydrogen embrittlement are being addressed. Alternate delivery forms such as the transport of hydrogen in safe compounds or chemical forms, are being developed to get hydrogen to end-use sites on an as-needed and real time usage basis.

Hydrogen is also distributed via cylinders and tube trailers that are transported by trucks, railcars, and barges. For long-distance distribution of up to 1000 miles, hydrogen is usually transported as a liquid and then vaporized for use on-site. Eleven plants have the capacity to produce 283 tons of liquid hydrogen per day in North America.

Hydrogen Delivery Methods

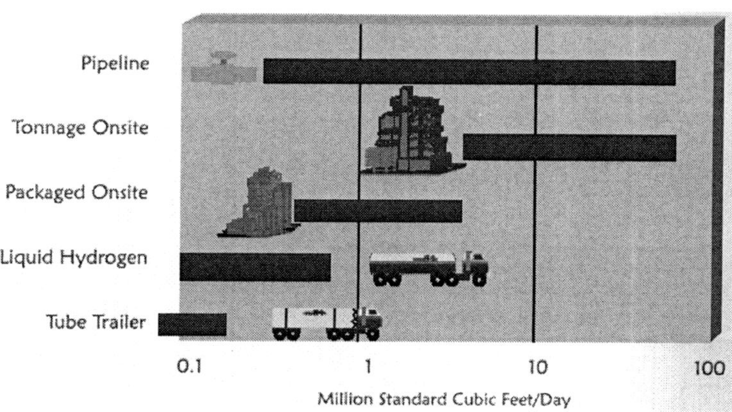

Source: Air Products and Chemicals, Inc.

Storage

Hydrogen can be stored as a gas or liquid or in a chemical compound using a variety of technologies.

Compact storage of hydrogen gas in tanks is the most mature storage technology, but is difficult because hydrogen is the lightest element and has very low density under normal conditions. This is addressed through compression to higher pressures or interaction with other compounds. In addition, storage tank materials are advancing—they are getting lighter and better able to provide containment. Some have a protective outside layer to improve impact resistance and safety.

Liquid hydrogen is stored in cryogenic containers, which requires less volume than gas storage. However, the liquefaction of hydrogen consumes large quantities of electric power, equivalent to about one-third the energy value of the hydrogen.

Hydrogen Storage Alternatives
Compressed Fuel Storage
• Cylindrical tanks
• Quasi-conformable tanks
Liquid Hydrogen Storage
• Cylindrical tanks
• Elliptical tanks
Solid State Conformable Storage
• Hydride storage material
• Carbon adsorption
• Glass microspheres

Source: QUANTUM Technologies

Hydrogen can be stored "reversibly" and "irreversibly" in metal hydrides. In reversible storage, metals are generally alloyed to optimize both the system weight and the temperature at which the hydrogen can be recovered. When the hydrogen needs to be used, it is released from the hydride under certain temperature and pressure conditions, and the alloy is restored to its previous state. In irreversible storage, the material undergoes a chemical reaction with another substance, such as water, that releases the hydrogen from the hydride. The byproduct is not reconverted to a hydride.

Laboratory research continues in the development of carbon-based storage systems. Hydrogen storage in carbon structures is achieved chemically in fullerenes or by physical sorption in carbon nanotubes. These processes are controlled through temperature and pressure and are still a long way from development.

Conversion

As mentioned, hydrogen is an energy carrier that requires production by an energy source (e.g., fossil, renewable, or nuclear) using a feedstock (e.g., fossil, biomass, or water) followed by consumption of the hydrogen by a particular end-use device to produce heat or electricity. Hydrogen can be converted to energy via traditional combustion methods and through electrochemical processes in fuel cells.

Combustion

Hydrogen can be combusted in the same manner as gasoline or natural gas. The benefit of using hydrogen combustion over fossil fuel combustion is that it releases fewer emissions—water is the only major byproduct. No carbon dioxide is emitted, and nitrogen oxides, produced by a reaction with the nitrogen in the air, can be significantly lower than with the combustion of fossil fuels. This technology is fairly well developed—both the National Aeronotics and Space Administration (NASA) and the Department of Defense use it for applications such as the space shuttle's main engines and unmanned rocket engines. Other combustion applications are being researched, such as new designs of combustion equipment specifically for hydrogen in turbines and engines. Hydrogen internal combustion engine vehicles are being demonstrated—Ford and BMW made significant progress in advanced Hydrogen Internal Combustion Engine (H2-ICE) vehicles in 2001. Also, the combustion of hydrogen blends is being practiced, as blends have been shown to emit fewer pollutants than pure fossil fuels once the engine is leaned out, and would promote a transition to the combustion of 100 percent hydrogen fuel.

Fuel Cells

Fuel cells utilize the chemical energy of hydrogen to produce electricity and thermal energy. A fuel cell is a quiet, clean source of energy. Water is the only by-product it emits if it uses hydrogen directly. Since electrochemical reactions generate energy more efficiently than combustion, fuel cells can achieve higher efficiencies than internal combustion engines. Current fuel cell efficiencies are in the 40 to 50 percent range, with up to 80 percent efficiency reported when used in combined heat and power applications.

Fuel cells are similar to batteries in that they are composed of positive and negative electrodes with an electrolyte or membrane. The difference between fuel cells and batteries is that energy is not recharged and stored in fuel cells as it is in batteries. Fuel cells receive their energy from the hydrogen or similar fuel that is supplied to them. No charge is thereby necessary.

Fuel cells are characterized by their electrolyte, operating temperature, and level of hydrogen purity required. The following table summarizes the characteristics of various fuel cell types. Phosphoric acid fuel cells are the most developed fuel cells for commercial use. Many of the installed units are used in stationary applications to provide grid support and reliable back-up power, and in transportation applications to power large vehicles such as buses. Proton exchange membrane (PEM) fuel cells are being developed and tested for use in transportation, stationary, and portable applications. There has been a tremendous upsurge in interest in PEM fuel cells over the past few years, and most major automotive manufacturers are developing fuel cell concept cars. Alkaline fuel cells have been used in military applications, for NASA space missions to provide electricity and drinking water for astronauts, and are being tested for transportation applications. Solid oxide and molten carbonate fuel cells are best for use in generating electricity in stationary combined cycle applications and cogeneration applications in which waste heat is used for cogeneration. They also fit well for portable power and transportation applications, especially large trucks.

Summary of Fuel Cell Types

Fuel Cell	Electrolyte	Operating Temperature (°C)	Sensitivities to Hydrogen Purity
Proton Exchange Membrane	Solid organic polymer poly-perfluorosulfonic acid	60-100	High sensitivities to impurities, must have <10 ppm CO
Alkaline	Aqueous solution of potassium hydroxide soaked in a matrix	90-100	High sensitivity to carbon dioxide
Phophoric Acid	Liquid phosphoric acid soaked in a matrix	175-200	Sensitive to CO
Molten Carbonate	Liquid solution of lithium, sodium and/or potassium carbonates, soaked in a matrix	600-1000	Low sensitivity to CO, Hydrogen/carbon monoxide mixtures can be used. is CO_2 required
Solid Oxide	Solid zirconium oxide to which a small amount of ytrria is added	600-1000	Low sensitivity to CO, Hydrogen/carbon dioxide/ methane mixtures can be used

Fuel cells have operating advantages for both stationary and mobile applications in that they are quiet and typically have high efficiencies at partial loads. They also have environmental advantages. For example, when pure hydrogen is used as the fuel, there are no emissions of sulphur or nitrogen oxides, or particulates. And if the hydrogen comes from a net-carbon-free renewable or nuclear energy source, the system will also be free of carbon dioxide emissions. The direct conversion of the energy stored in the fuel to electricity in a fuel cell can be achieved at high efficiencies, avoiding limitations of standard heat-to-power cycles used in combustion engines and turbines. Fuel cells are also deployable in combined heat and power applications.

End Use Energy Applications

Hydrogen energy end-use applications include stationary, transportation, and portable devices. As mentioned, the most common current use for hydrogen is in industrial processes such as refineries. It is also used as a fuel at NASA, where the combustion of hydrogen has fueled its space shuttle main engines and propulsion systems for years. Other energy uses are generally limited to research and demonstrations.

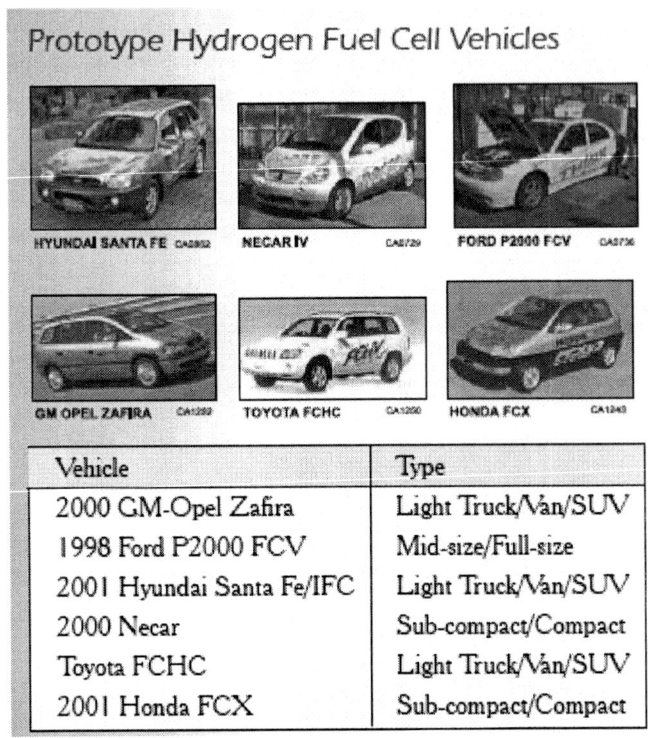

Vehicle	Type
2000 GM-Opel Zafira	Light Truck/Van/SUV
1998 Ford P2000 FCV	Mid-size/Full-size
2001 Hyundai Santa Fe/IFC	Light Truck/Van/SUV
2000 Necar	Sub-compact/Compact
Toyota FCHC	Light Truck/Van/SUV
2001 Honda FCX	Sub-compact/Compact

Source: UTC Fuel Cells

One application of hydrogen fuel cells is for distributed generation. A number of UTC Fuel Cell's phosphoric acid fuel cells are operating in locations around the world, providing heat and power for buildings and industrial applications. These units include a reformer component to generate a hydrogen-rich gas from natural gas.

In the transportation sector, a number of fuel cell vehicles are being tested and developed. Vehicular use of hydrogen energy requires a compact power system and refueling stations. Given the current state of hydrogen technologies, city-owned buses are a promising application because they are capable of carrying large tanks of hydrogen and typically refuel at a single location. In March 1998, for example, Chicago became the first city in the United States to use hydrogen fuel cells to power buses in their public transit system.

Several car manufacturers, including Hyundai, Ford, General Motors, Toyota, Honda, and DaimlerChrysler are developing fuel cell vehicles for personal use. In 2001, the BMW CleanEnergy World Tour demonstrated its fleet of 15 hydrogen-powered internal combustion engine vehicles.

Portable fuel cells can also be used to power small devices such as mobile telephones or personal computers. Larger power generators for recreation and other off-grid applications are under development. For example, Ballard Power Systems has developed the Nexa™ power module, a PEM fuel cell system that generates up to 1200 watts of unregulated direct current electrical power that can be used for industrial and consumer end-product applications. This portable power application is still under development. Today's emerging hydrogen energy industry is eager to develop hydrogen fuel infrastructure technology that can be used to generate power for stationary, transportation, and portable power applications. Much work needs to be done to reach this goal, but a foundation for future efforts has been established by these various technology sectors.

3. KEY DRIVERS AFFECTING THE FUTURE OF HYDROGEN ENERGY DEVELOPMENT

The United States' energy sector is experiencing a confluence of events. New technologies are being developed and opportunities for entrepreneurial ideas and innovative approaches are ripening at a time when our capital-intensive, aging energy infrastructure is in need of improvement. Despite this window of opportunity, the overall business environment for energy investments in America today is not conducive to the massive introduction of new technologies.

The Nation faces uncertainties in our energy future and inertia in our infrastructure system. America's energy future will include unpredictable ups and downs, price volatility, regional gluts and shortages, and market instabilities. The natural pace of turnover of existing capital in our infrastructure is relatively slow, there is reluctance to alter traditional systems, and the framework of changing policies and regulations tends to favor incumbent suppliers and technologies.

> The Nation faces uncertainties in our energy future and inertia in our infrastructure system.

These factors introduce uncertainties and risk and interfere with making changes. For example, existing inertia in our energy system has made it difficult for policy makers and business executives to make strategic decisions about long-term energy requirements, which has led to delays in decision-making, and has made it hard for businesses to commit to large financial resources to energy investments.

The factors affecting hydrogen's potential are rooted in these issues. Our infrastructure has been designed to provide users with reliable supplies of fossil fuels at an affordable cost while protecting the environment. Other forms of energy, including nuclear and renewable sources, may play important roles but face their own hurdles in the global competition for market share.

Summary of Key Drivers Affecting Hydrogen Energy Development

Support	Inhibit	Both Support and Inhibit
National security and the need to reduce oil imports	The inability to build and sustain national consensus on energy policy priorities	Rapid pace of technological change in hydrogen and competing energy sources and technologies
Global climate change and the need to reduce greenhouse gas emissions and pollution	Lack of a hydrogen infrastructure and the substantial costs of building one	The current availability of relatively low-cost fossil fuels, along with the inevitable depletion of these resources
Global population and economic growth and the need for new clean energy supplies at affordable prices, as hydrogen is potentially available in virtually unlim-ited quantities	Lack of commercially available, low-cost hydrogen production, storage and conversion devices, such as fuel cells Hydrogen safety issues	Simultaneous consumer preferences for both a clean environment and affordable energy supplies
Air quality and the need to reduce emissions from vehicles and power plants		

In developing a vision for the hydrogen economy, certain questions about market and policy forces arise. Are there technological, economic, or policy-related factors, issues, or trends that can encourage a dynamic of change in our current market? If so, how might they affect hydrogen energy development over the next several decades, in terms of supporting or hindering it. Many of these "drivers" will affect not only hydrogen, but also the future of the energy system as a whole.

The tragic events of September 11, 2001, remind every American of the danger of reliance on oil imports from politically unstable countries.

National Security

The need to enhance the supply of domestically produced transportation fuels is great. The tragic events of September 11, 2001, remind every American of the danger of reliance on oil imports from politically unstable countries, some of which have opposing interests to those of the United States. America's transportation sector relies almost exclusively on refined petroleum products; more than one-half of the petroleum consumed in the United States is imported, and that percentage is expected to rise steadily for the foreseeable future, unless we change our energy use. Hydrogen (along with biofuels) is a versatile energy carrier that could be produced entirely from domestic sources of fossil fuel (e.g., natural gas and coal with capture and sequestration of carbon dioxide), renewable (e.g., solar, wind, and biomass), and nuclear energy, in large quantities. Its use as a major energy carrier would provide the United States with a more diversified energy infrastructure.

Climate Change

The combustion of fossil fuels accounts for the majority of anthropogenic greenhouse gas emissions released into the atmosphere. Although international efforts to address global climate change have not yet resulted in policies that all nations have accepted, there is growing recognition that steps to reduce greenhouse gases are needed, and many countries are adopting policies to accomplish that end. Energy and transportation companies, many of which have multi-national operations, are actively evaluating alternative sources of energy.

Hydrogen can play an important role in a low-carbon global economy, as its only byproduct is water. With the capture and sequestration of carbon from fossil fuels, hydrogen is one path for coal, oil, and natural gas to remain viable energy resources, should strong constraints on carbon emissions be required. Hydrogen produced from renewable resources or nuclear energy results in no net carbon emissions.

Population and Economic Growth

Many experts have pointed out that if highly populated countries like China, India, or Indonesia were to adopt energy consumption patterns similar to those of the United States or Europe, world energy supplies would have to increase enormously to meet demand. For example, imagine the energy and environmental consequences of one-half of Chinese households owning automobiles that run on gasoline. That would be about four times the number of car-owning households in the United States today. If these automobiles were to run on hydrogen fuel cells, the environmental consequences and national security issues would be much less.

Many international energy experts are hoping that developing countries will be able to "leap frog" today's energy devices and infrastructure by adopting advanced technologies. The idea is that as economic growth spreads around the world, developing countries would be able to follow a pattern similar to the one being followed in telecommunications systems: wireless technologies are being installed in certain locations, "leap frogging" the need for telephone lines. Unfortunately, the advanced energy devices that would be needed to "leap frog" current infrastructure, such as hydrogen energy systems and fuel cells, are not yet cost competitive or commercially available on the required economies of scale.

Air Quality

Air quality is a major public health concern. Most of the major metropolitan areas in the United States are in "non-attainment" with the requirements of the Clean Air Act. States are required to develop strategies detailing the steps they plan to take for reaching national ambient air quality goals. California, for example, has been one of the most aggressive states in taking action to reach clean air goals. Personal vehicles and electric power plants are significant contributors to the nation's air quality problems. The introduction of hydrogen-using commercial bus fleets is one of the approaches being taken to achieve compliance with the Clean Air Act.

Without such policies, progress toward a hydrogen economy will be severely limited and much slower.

Consistency in National Energy Policy Priorities

One of the hallmarks of public energy policies in America is their inconsistency from state to state, and the shifts that occur in Federal priorities with each election. The last major piece of Federal energy legislation was the Energy Policy Act, which was enacted in 1992. Since then, there has not been a national consensus on Federal energy policy priorities. For example, prior to the electricity problems experienced in California in 2000 and 2001, electricity restructuring was one of the top national energy priorities. The last three Congresses have held hearings on this subject and there have been Administration electricity legislation proposals, but there has not been a long-term, consistent energy policy.

The United States is missing a sustained national commitment to environmental and energy security goals, and the policies to support them. Hydrogen could provide the basis for such a policy. Since 1990 Congress has authorized funds in support of hydrogen energy research, development, and demonstration. Market-based environmental policies that provide industries with financial reasons to invest in low-emission or carbon-free energy systems could accelerate hydrogen energy development substantially.

The public needs to understand the value of a hydrogen economy in order for businesses to invest in new energy technologies. Public policies need to be developed by government and private entities and put into place to facilitate public acceptance. This in turn would lead to greater market incentives for significant private investment in hydrogen.

Hydrogen Infrastructure Costs

As documented in the President's National Energy Policy, America's energy infrastructure is aging and in need of significant upgrades, overhauls, and replacements over the next several decades. This infrastructure includes oil refineries, gas and oil pipelines, power plants, and electricity transmission and distribution facilities such as power lines, transformers, and substations. The capital investment requirements to maintain and improve the infrastructure over the next several decades will total hundreds of billions of dollars.

While hydrogen may be able to use some of the existing infrastructure, specific upgrades and enhancements will be needed to accommodate the unique features of hydrogen, particularly in storage and distribution. The technologies needed to convert the natural gas infrastructure for the use of hydrogen are available today, but are not yet cost-effective. At present there is no motivation to convert to hydrogen, as there are essentially no markets for distributed use of hydrogen energy. Additional infrastructure costs will have to be incurred in the future, when cost-competitive products are available, to enable the transition to the hydrogen economy.

The technical and economic barriers to upgrading the Nation's fueling stations to provide hydrogen represents one of the major stumbling blocks to the expanded use of hydrogen-fueled vehicles. Some automakers estimate that hydrogen would have to be available in at least thirty percent of the nation's fueling stations for a viable hydrogen-based transportation sector to emerge. Private investment in such an infrastructure will not be forthcoming in the absence of supporting and sustained, supportive public policies.

For the hydrogen economy to evolve, consumers will need to have convenient access to hydrogen sources.

Hydrogen Storage and Conversion Devices

The lack of low-cost and light-weight storage and commercially-available and cost-competitive fuel cells interferes with the development of a hydrogen economy. For the hydrogen economy to evolve, consumers will need to have convenient access to hydrogen, and storage will be one of the keys. Better hydrogen storage systems will enable users to have easy access to hydrogen for vehicles and distributed energy facilities. Hydrogen storage will also enhance the value and potential market share of renewable electricity generation.

Fuel cells are clean, compact, and modular energy generation devices that have the potential to revolutionize the production of electricity and thermal energy, for both stationary and mobile applications. There are several different types of fuel cells; each has advantages and disadvantages. Design and manufacturing breakthroughs are needed to lower costs and enhance reliability and performance. The marketplace will determine which of the several fuel cell options will offer users the most favorable advantages.

Hydrogen internal combustion engine vehicles are being demonstrated and are nearer term than fuel cell vehicles. They offer nearly as many benefits, although they are not strictly zero emissions vehicles.

Concerns About Hydrogen Safety

Perceptions about the safety of hydrogen remain a deterrent to many consumers. The public needs to be aware that safety issues related to hydrogen are being addressed, and perceptions based on misinformation need to be corrected. A public information campaign can help eliminate many of the concerns about hydrogen safety. Effective codes and standards are needed to ensure that these concerns are addressed in equipment designs, manufacturing practices, and operation and maintenance procedures.

Appropriate field tests and demonstrations will be needed to increase public confidence and acceptance of hydrogen technologies.

Technological Change

The accomplishments of science over the past century have been astounding. America's energy sector has been a major beneficiary of developments in science and engineering. The electrification of the economy and the ease of personal travel through the use of the automobile are part of the reason why the American economy has become one of the most productive in history.

Examples of the accelerating pace of technological change in recent years are everywhere. There have been many innovations and new product and service offerings in telecommunications, information systems, and biology-based industries. As a result, these have become leading sectors in the United States economy. Development of advanced energy systems, products, and services are also on the horizon. New materials for energy devices such as turbines, engines, and fuel cells are improving rapidly. Advanced concepts in the biological sciences, and in the miniaturization of engineered systems using nanotechnologies, could lead to breakthroughs that revolutionize how energy (including that from hydrogen sources) is produced and used.

At the same time, these developments are also affecting conventional fossil, nuclear, and renewable energy systems, which are becoming cleaner, cheaper, and more energy-efficient. For example, recent advances in combined heat and power systems have led to systems that produce electricity and thermal energy at more than double the energy efficiency of the typical power plant, thus reducing emissions, energy costs, and peak electricity demands.

Americans demonstrate preferences that both support and hinder the development of clean energy sources.

Availability of Fossil Energy Resources

Affordable coal, oil, and natural gas supplies are available around the world. Analysts warn, however, that world oil production cannot be sustained at current levels indefinitely and that the development of America's natural gas resources, while extensive, are not inexhaustible. Coal is the major fuel for electricity production in America. Clean coal technologies improve efficiency and reduce emissions. Fossil fuels are expected to be America's fuels of choice for the foreseeable future.

If demand for fossil fuels continues to increase, resource constraints will push fossil fuel prices up over the next several decades (energy analysts differ as to how much and when). This will spur the development of non-fossil alternatives such as solar, wind, geothermal, biomass, and nuclear, and fossil alternatives that sequester the carbon dioxide, and encourage the transition to hydrogen. In the meantime, the current availability and relatively low cost of fossil fuels moderate the pace of development of alternative sources of energy.

Consumer Preferences

Americans demonstrate preferences that both support and hinder the development of clean energy sources. The typical consumer has grown increasingly aware of the environmental, health, and safety consequences of energy. This, in combination with legislation, has contributed to the growth in demand for environmentally friendly products and services such as energy efficient appliances and equipment, renewable energy, and "green power" offerings from local utilities. In many non-attainment areas, for example, environmental requirements to reduce emissions receive public support, even when they result in higher fuel taxes or inconvenience in commuting.

There are also other energy consumer trends that require high-value services that command premium prices. For example, extra high reliable electric service is required by a growing number of businesses relying on digital equipment, such as computers, or operating 24 hours a day, seven days a week. Protection against energy price volatility is another possible high-value service. These premium markets, like distributed energy and power parks, provide growing opportunities for hydrogen energy, especially early in the transition.

In contrast, public support for low energy prices is strong. Americans enjoy gasoline prices that are among the lowest in the world. Consumers rank fuel economy relatively low on the list of desired attributes for automobiles. The ten-year period from 1992 to 2001, the longest era of sustained economic growth in American history, coincided with a period of historically low energy prices. Many Americans are quick to contrast that era with the ten-year period from 1972 to 1981—an era marked by relatively low economic growth, high inflation and unemployment, and high energy prices.

Summary

There are several key drivers that will likely affect hydrogen energy development. Concerns regarding national security, global climate change, and worldwide population and economic growth will increasingly promote systems that support hydrogen development. The lack of a national consensus on energy policy priorities, a hydrogen infrastructure, commercially viable hydrogen technologies, and the public perception of hydrogen safety issues have the potential to inhibit hydrogen energy development. Drivers that could both support and inhibit the development of hydrogen are the rapid pace of technological change in energy technologies, the current availability of low cost fossil fuels and their eventual depletion, and mixed consumer preferences for clean and cheap energy.

4. A National Vision of the
U.S. Hydrogen Economy

Vision: Hydrogen is America's clean energy choice. It is flexible, affordable, safe, domestically produced, used in all sectors of the economy, and in all regions of the country.

At the time this vision is realized, hydrogen produced from fossil fuels (with carbon capture and sequestration), renewable energy, and nuclear energy will be used throughout the transportatiion and electric power sectors. Hydrogen will be produced in centralized facilities in remote locations, in power parks and fueling stations in our communities, in distributed facilities in rural areas, and onsite at customers' premises. Today we have a hydrocarbon economy. Tomorrow we will have weaned ourselves from carbon and will live in a "hydrogen economy." In the hydrogen economy...

America will enjoy a secure, clean, and prosperous energy sector that will continue for generations to come. American consumers will have access to hydrogen energy to the same extent that they have access to gasoline, natural gas, and electricity today. It will be produced cleanly, with near-zero net carbon emissions, and it will be transported and used safely. It will be the "fuel of choice" for American businesses and consumers. America's hydrogen energy industries will be the world's leaders in hydrogen-related equipment, products, and services.

One major foundation for this vision is the development of an energy infrastructure that can support the expanded production, delivery, storage, and use of hydrogen energy. A hydrogen infrastructure is likely to develop on a regional level, since it can be made from a variety of feedstocks, and will resemble the electricity grid more than the current oil delivery system. Construction of this infrastructure will take time and will require significant resources. As a result, the hydrogen economy will evolve over the next several decades. Hydrogen storage weight and volume reductions, mass production of fuel cells, construction of the necessary infrastructure, and expanded use of portable and distributed power generation devices will sustain the momentum towards a hydrogen economy. Infrastructure will begin with pilot projects and expand to local, regional, and ultimately national and international applications.

A strong public-private partnership will be a key feature of this evolutionary process. Together, government and private organizations will facilitate appropriate research, development, and demonstration programs; educate the public; and develop codes and standards. Mounting public pressures for a cleaner environment, and for more secure sources of energy supplies, will lead to more stringent policies for reducing air emissions and limiting oil imports.

With steady progress and a few significant technology breakthroughs, the Nation will make a committed switch to a hydrogen economy—over the next several decades a confluence of events will mark a steep increase in hydrogen energy development. By that time, hydrogen production costs will be lower, the basic components of a national hydrogen storage and distribution network will be in place, and hydrogen-powered fuel cells, engines, and turbines will be mature technologies that are mass produced for use in cars, homes, offices, and factories.

Early glimpses of this vision can already be seen in pilot programs that are underway in a few U.S. locations and several other countries; Iceland is a notable example. There are many advantages in partnering with other countries in the development of new technologies for the "hydrogen economy."

> **Icelandic New Energy Ltd.**, a joint venture company endorsed by Iceland's government, was formed in 2000 to work towards a hydrogen economy. It leads new projects that facilitate the use of hydrogen as an alternative fuel to reduce greenhouse gas emissions, increase energy efficiency, and protect depleting natural resources. Icelandic New Energy plans to gradually switch the nation's vehicles—first buses, then cars, then fishing vessels—to hydrogen power. The company expects to have three hydrogen-powered buses running in Reykjavik by the end of 2002, fuel cell-powered passenger cars available in 2003 to 2004, and a demonstration of a fuel cell boat in 2006.

FUTURE HYDROGEN PRODUCTION PROCESSES

At the time the vision for a hydrogen economy becomes a reality, several decades from now, hydrogen will still be produced from fossil fuels, but also from biomass and water using thermal, electric, and photolytic processes.

Hydrogen produced from water will be a cost competitive alternative to hydrogen made from hydrocarbons. Some techniques will include inexpensive electrolyzers that use solar, geothermal, wind, or nuclear power; photochemical or photoelectrochemical devices; and biological systems such as algae. Very low cost electricity will be needed for costcompetitiveness, such as off-peak power at one to two cents per kilowatt hour, and the amount of hydrogen produced in this manner will be limited in comparison to future demands for fuel. Electrolysis could make locally important contributions where low cost power is available, although the overall contribution to a full scale hydrogen economy will depend on the relative costs.

The Nation will have a combination of central station and distributed hydrogen production facilities; the mix will depend on local economics and regional resource endowments. Central station facilities will consist of multiproduct refineries that use fossil fuels or biomass as feedstocks, and that provide hydrogen, electricity, thermal energy, chemicals, and other industrial products. One can also envision central station nuclear, solar, wind, or geothermal facilities for the production of hydrogen by splitting water. Those facilities that use fossil fuels and biomass as feedstocks will have carbon capture and sequestration capabilities. Distributed production of hydrogen will occur on-site (at homes, offices, factories, and onboard vehicles) or at the community level at places such as fueling stations or power parks.

Sources of Hydrogen

Conversion Process (Today → Future)

Sources	Thermal	Electric	Photolytic
Fossil Fuels	✓		
Biomass	✓		✓
Water		✓	✓

Future Infrastructure

A national network will be in place to provide hydrogen to users in every region, state, and locality. The network will evolve from the existing fossil fuel-based infrastructure and will accommodate both centralized and decentralized production facilities. Pipelines will be the preferred choice for distributing hydrogen to high-demand areas. Trucks and rail will be used to distribute hydrogen to rural and other lower-demand areas. On-site hydrogen production and distribution facilities will be available where demand is high enough to sustain maintenance of the technologies.

Future Storage Devices

A selection of relatively lightweight, low cost, and low volume hydrogen storage devices will be available to meet a variety of needs. Pocket-sized containers will provide hydrogen for portable telecommunications and computer equipment, small and medium hydrogen containers will be available for vehicles and on-site power systems, and industrial-sized storage devices will be available for power parks and utility-scale systems. Solid-state storage media that use metal hydrides will be mature technologies in mass production. Storage devices based on carbon structures will be under development.

Future Conversion Technologies

Fuel cells will be mass-produced and will be cost-competitive and mature technologies. Advanced hydrogen-powered energy generation devices such as combustion turbines and reciprocating engines will be in widespread commercial use.

Future End-Use Energy Markets

Hydrogen will be available for every end-use energy need in the economy, including transportation, power generation, and portable power systems. Hydrogen will be the dominant fuel for government and commercial vehicle fleets. It will be used in a large number of personal vehicles and light duty trucks. It will be combusted directly and mixed with natural gas in turbines and reciprocating engines for electricity and thermal energy in homes, offices, and factories. It will be used in fuel cells for both mobile and stationary applications. And it will be used in portable devices such as computers, mobile phones, Internet hook-ups, and other electronic equipment.

This chapter presented a vision of the hydrogen economy. The following chapter discusses in further detail how the United States can transition to a hydrogen economy through public-private partnerships.

> **SunLine Transit Agency** opened a hydro-
> gen generation, storage, and fueling facility in
> April 2000 in Thousand Palms, California,
> with the help of industry and government
> partners. SunLine uses electrolyzers to
> generate hydrogen from renewable energy
> (photovoltaics) and a reformer to generate it
> from natural gas. The agency stores hydrogen
> in a 16-tube storage trailer that is attached to a
> cascade control panel used to fill hydrogen
> buses and pickup trucks at a public fueling
> island. SunLine plans to add another
> electrolyzer, expanded storage, a wind-gener-
> ated fuel cell demonstration project, and
> increased dispensing pressure to its "Clean
> Fuels Mall" in the near future.

5. Transition Ransition to the Hydrogen Economy

There are certain achievements to be made and pitfalls to avoid in the transition to a hydrogen economy. Rather than offering predictions or specific prescriptions, the following section discusses various goals and hazards and outlines plausible scenarios.

Phase One – Progres in Technologies, Policies, and Markets

Significant laboratory progress is expected in the first transition phase in the form of research and demonstration that supports industry's pre-commercial efforts. Carmakers will test several types of hydrogen-using prototypes. Research will focus on bringing down the cost of fuel cells and developing solid-state storage devices, primarily using metal hydrides, but also exploring carbon structures (such as nanotubes and fullerenes) and glass microspheres.

Natural gas steam reforming will continue to be the primary means for producing hydrogen. While progress will be made in developing advanced hydrogen production technologies, market readiness will not yet be achieved.

Hydrogen use in internal combustion engines will build support for infrastructure development. Testing of fuel cell buses and cars and of hydrogen-fueled internal combustion engines will likely continue to expand, particularly in non-attainment areas. Industry markets for proton exchange membrane fuel cells for stationary applications will be further developed, and auto companies will begin to build hydrogen fuel cell cars in quantity. Tests of fuel cells for combined heat and power applications in buildings will increase, and development of hydrogen fuel cells for portable power devices will continue.

The first phase will also include the creation of hydrogen-related policies on energy and the environment, including the reduction of energy imports, managing greenhouse gas emissions, and strengthening the control of air pollution. International standards for the safe use of hydrogen will be implemented around the world. The restructuring of electricity and natural gas markets in the United States will be completed, thus expanding the prospects for the installation of distributed energy systems. Government will address liability, permitting, and codes and standards in order to provide a framework for commercial development to proceed.

> *The transition to the hydrogen economy has already begun. We are a hydrocarbon economy today. We have yet to learn how to inexpensively extract and convert hydrogen from hydrocarbon (or other) sources without releasing harmful by-products into the environment. But we are quickly learning....*

Phase Two – Transitioning to the Marketplace

Many significant technology developments will have to occur in the next phase in order for the hydrogen economy to develop. The most necessary breakthrough will have to be cost reductions of fuel cells through the development of large-scale manufacturing capabilities for stationary and mobile units. The initial stages of a hydrogen delivery system will have to be in place.

Specifically, significant advances will have to be made to lower the cost of hydrogen production and storage. Natural gas reforming will remain the primary source of hydrogen production. Coal gasification and the use of nuclear and renewable technologies will be used more heavily. Lighter-weight and lower-cost storage devices will become commercially available.

Power parks and fueling stations will include distributed hydrogen production systems, some of which will be renewable (e.g., solar and wind). A few will use photobiological and photoelectrochemical techniques. Hybrid hydrogen internal combustion engines will be further developed and more widely used.

Federal and state government facilities will play an increasingly expanded role in moving hydrogen technologies into the marketplace. Many will have served as "first use" sites for hydrogen energy systems, such as municipal bus services. Urban emergency services, fire, and police facilities will also use distributed energy devices to ensure continuous power, and some will certainly opt for hydrogen-based systems because of the environmental advantages. Military applications of hydrogen energy systems (e.g., in vehicles, ships, and aircraft) will be demonstrated, and there will be numerous instances of hydrogen fuel cells providing combined heat and power services for buildings.

Phase Three – Expansion of Markets and Infrastructure

This is the phase in which technology advancements in hydrogen extraction will reduce costs and expand market share.

The market will be penetrated with the widespread use of fuel cell-powered buses and government vehicles. Emphasis will be placed on expanding from local pockets of hydrogen energy development to building a national hydrogen infrastructure. Although hydrogen production will often occur onsite or onboard, it will also be produced in large-scale refineries that will use coal or biomass as a feedstock for the simultaneous production of hydrogen, electricity, thermal energy, chemicals, and other fuels.

Large- and small-scale hydrogen storage using hydrides will become mature technologies and enter mass production. Other advanced storage techniques such as carbon structures will be in the development stage but close to commercialization. National policies will support hydrogen market expansion, and state and local standards will be in place. Siting and permitting of hydrogen technologies will be more streamlined. State and local governments will play a major role in this phase.

Phase Four – Realization of the Hydrogen Vision

Eventually hydrogen will overtake fossil fuels for most end-use energy market applications. Economical and environmentally friendly means will be found for extracting hydrogen from fossil fuels, biomass, and water. Hydrogen "farms" that use biological systems such as algae to extract hydrogen from water will be in use. Gasification plants will extract hydrogen from coal and biomass. Carbon capture will limit emissions, and new industrial uses will put captured carbon to work for industrial feed stocks, building materials, and other applications.

A national infrastructure that supports the use of hydrogen for fuel and electricity production will be in place. United States companies that spent decades developing hydrogen technologies will be exporting products and services around the world. American consumers will be enjoying the economic benefits of a financially sound hydrogen energy sector and the environmental benefits of clean energy systems.

The market for hydrogen vehicles as personal transportation will expand as a natural outgrowth of technology, market, and policy development. There will be no need for government mandates. The practice of using hydrogen vehicles to provide heat and power for the workplace during the day, and homes during the night, will be commonplace. The line between the transportation sector and the power system will blur. The hydrogen economy will become reality.

6. A PATH FORWARD

If significant progress toward a hydrogen economy is to occur, a new public-private partnership needs to form and spring into action immediately.

The ultimate vision for the hydrogen economy is decades in the future and the amount of research, development, public education, institution building, and infrastructure construction needed to get there is enormous.

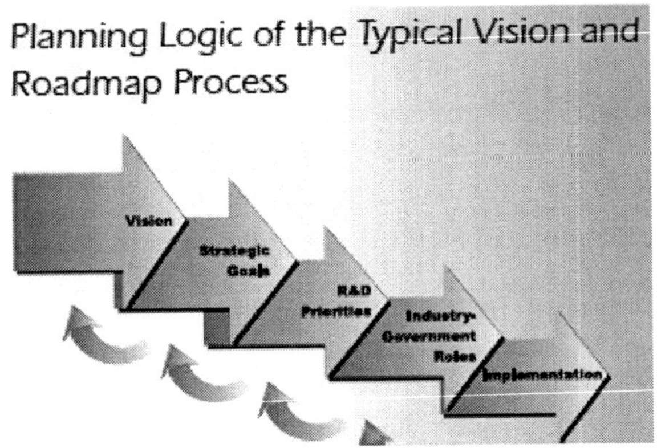

A useful venue for building this new public-private partnership is the joint development of a *National Hydrogen Energy Roadmap*. The roadmapping process can provide industry practitioners, government officials, and technologists from universities and the National Laboratories with the opportunity to collectively identify near-, mid-, and long-term actions. The process can be used to set priorities for research, development, and demonstration programs, and it can outline the relative roles of industry, government, universities, National Laboratories, and non-governmental organizations. This roadmap process will need to address a number of areas:

- Technologies for hydrogen production
- Technologies for hydrogen delivery and transportation
- Technologies for hydrogen storage
- Technologies for hydrogen conversion
- Scope and directions for public-private partnerships, both nationally and internationally
- Codes and standards for safe production, delivery, and use of hydrogen
- Education of the general public and government and private decision makers about the potential benefits from the expanded use of hydrogen

- End-use energy markets for hydrogen including the potential for "first use" fleet applications in Federal facilities, vehicles, and equipment

Participants in the roadmap process should include technical experts and industry practitioners from domestic and international organizations, as they have the requisite capabilities and experience to chart the critical hydrogen energy development pathways over the next several decades.

Examples of possible roadmap participants are listed in the table below.

Roadmap Areas	Roadmap Participants
Technology development	Industry, universities, National Laboratories
Market applications	Equipment manufacturers, industrial end users, building owners and operators, and Federal energy managers
Public education	Federal, State, and local governments; school districts; universities; media companies
Codes and standards	State and local governments, professional associations, standards organizations

The roadmap needs to be systematic from resource to end-use. Interfaces matter. For example, it may not be best to have refueling stations at one pressure and onboard storage tanks at another.

Working together, industry, universities, and government can help America realize the hydrogen vision. This entails building on our existing energy infrastructure and current hydrogen energy technologies to meet mutually set milestones. Public and private entities can cooperate to overcome hurdles and develop technologies and policies that fit America's drive for clean, affordable, secure, and efficient energy systems.

APPENDIX: PARTICIPATING ORGANIZATIONS

Air Products and Chemicals, Inc., Arthur Katsaros

Alameda Contra Costa Transit, Jaimie Levin

Avista Labs, Inc., J. Michael Davis

Ballard Power Systems, Stephen Kukucha

Battelle-Pacific Northwest National Laboratory, Jae Edmonds

BP, Lauren Segal

California Energy Commission, Susan Brown and Louise Dunlap

DaimlerChrysler, William Craven

DCH Technology, Inc., John Donohue

Energetics, Inc., Rich Scheer, Jack Eisenhauer, Ross Brindle (Facilitators)

Entergy Nuclear, Inc., Dan Keuter

ExxonMobil Refining and Supply Company, William Lewis

Ford Motor Company, Frank Balog

GE Corporate Research and Development, Sanjay Correa

General Motors, Byron McCormick

Hawaii Natural Energy Institute, Richard Rocheleau

House Committee on Science, John Darnell

Kennedy Space Center, David Bartine

Millennium Cell, Stephen Tang

National Academy of Sciences, James Zucchetto

National Energy Technology Laboratory, Rita Bajura

National Governors Association, Ethan Brown

National Renewable Energy Laboratory, Richard Truly

Natural Resources Defense Council, Daniel Lashof

NiSource Inc., Arthur Smith

Office of Assistant Secretary of the Navy (IandE), Leo Grassilli

Office of Naval Research, Richard Carlin

Office of Senator Akaka, Jaffer Mohiuddin

Praxair, Inc., Donald Terry

Princeton University, Joan Ogden

Proton Energy Systems, Inc., William Smith

Quantum Technologies, Alan Niedzwiecki

Stuart Energy Systems, Andrew T.B. Stuart

SunLine Transit Agency, Richard Cromwell

Texaco Energy Systems, Graham Batcheler

The Wexler Group, Robert Walker

U.S. Department of Energy, Robert Card, Robert Dixon, David Garman, Tom Gross, R. Shane Johnson, Karen Kimball, Robert Kripowicz, Richard Moorer, William Parks, Frank Wilkins

U.S. House of Representatives, Congressman Roscoe Bartlett

UTC Fuel Cells, William Miller
Verizon, Thomas Bean

World Resources Institute, James MacKenzie

INDEX

D

recreation, 91
recycling, 36
reduction, 38, 45, 46, 105
refining, 14, 21, 81
regulations, 21, 56, 57, 65, 91
regulators, 58
reliability, 12, 26, 45, 46, 49, 50, 62, 96
renewable energy, 97, 98, 99
resistance, 34, 86
resources, 4, 7, 9, 10, 11, 14, 15, 22, 24, 56, 60, 61, 64, 83, 92, 93, 98, 100
restructuring, 95, 105
retail, 25
returns, 30
rights, 55
risk, 3, 10, 12, 40, 45, 58, 63, 91

S

safety, 12, 13, 15, 20, 27, 30, 34, 38, 42, 45, 46, 48, 53, 56, 58, 59, 62, 63, 86, 92, 96, 98, 99
sales, 25
scaling, 23
scientific understanding, 47
security, 1, 2, 10, 11, 13, 14, 15, 19, 52, 59, 62, 63, 76, 77, 92, 95
self, 38
Senate, 1
sensing, 58
sensitivity, 89
sensors, 12, 32, 33, 39, 47, 53, 62
separation, 83
September 11, 76, 92, 93
series, 81
services, 13, 44, 45, 64, 65, 97, 98, 100, 106, 107
severity, 57
shape, 80
sharing, 55
sites, 25, 27, 31, 48, 82, 85, 106
sodium, 89
solid state, 23
sorption, 87

speed, 13, 64
stages, 26, 43, 45, 52, 106
stakeholders, 11, 13, 14, 60, 61, 76, 80
standards, 4, 13, 16, 18, 20, 21, 31, 33, 45, 48, 53, 55, 57, 58, 64, 65, 79, 97, 100, 105, 107, 108
steel, 85
storage, 3, 4, 6, 11, 12, 14, 15, 16, 18, 19, 24, 26, 27, 32, 34, 35, 36, 37, 38, 39, 40, 41, 42, 44, 48, 50, 52, 53, 56, 59, 60, 62, 63, 64, 65, 77, 81, 82, 84, 86, 87, 92, 95, 96, 100, 103, 104, 106, 107, 108, 109
strategies, 52, 53, 94
strength, 34
students, 56, 57, 58, 60
sulfur, 81
sulphur, 89
summaries, 76
summer, 60
suppliers, 19, 91
supply, 3, 9, 12, 16, 18, 19, 22, 23, 29, 62, 83, 85, 93
surface area, 36
switching, 57
systems, 3, 10, 11, 12, 13, 17, 19, 23, 26, 28, 30, 32, 33, 34, 36, 39, 40, 41, 45, 46, 49, 50, 51, 53, 55, 56, 57, 59, 60, 62, 63, 64, 75, 76, 77, 79, 81, 82, 84, 87, 89, 91, 94, 95, 96, 97, 99, 103, 105, 106, 107, 109

T

tanks, 29, 34, 39, 42, 82, 86, 90, 109
targets, 33, 37, 50
tax credit, 55
taxonomy, 7
teacher training, 60
teachers, 58, 60
technological change, 92, 97, 99
technology, 1, 4, 5, 6, 7, 10, 12, 13, 14, 15, 19, 20, 22, 25, 26, 34, 36, 37, 38, 39, 40, 42, 43, 44, 45, 47, 50, 57, 58,